The
OXFORD
MATHEMATICS
STUDY
DICTIONARY

Frank Tapson

mathematics Very broadly, mathematics concerns itself with patterns and relationships within, and between, the all-inclusive ideas of number and space. It is divided into several branches; some of the better known are arithmetic, geometry, algebra, trigonometry, calculus and statistics. *Mathematics is used by many other subjects in an attempt to analyse them, and to impose order and organisation on them.*

OXFORD
UNIVERSITY PRESS

OXFORD
UNIVERSITY PRESS

Great Clarendon Street, Oxford OX2 6DP

Oxford University Press is a department of the University of Oxford. It furthers the
University's objective of excellence in research, scholarship, and education by
publishing worldwide in

Oxford New York
Athens Auckland Bangkok Bogotá Buenos Aires Calcutta
Cape Town Chennai Dar es Salaam Delhi Florence
Hong Kong Istanbul Karachi Kuala Lumpur Madrid
Melbourne Mexico City Mumbai Nairobi Paris
São Paulo Shanghai Singapore Taipei Tokyo Toronto

and associated companies in
Berlin Ibadan

Oxford is a registered trade mark of Oxford University Press in the UK
and in certain other countries

First published 1996
Second edition 1999

Reprint for this edition: 10 9 8 7 6 5 4
ISBN   0 19 914567 9

A CIP catalogue record for this book is available from the British Library.

Typeset and illustrated by the author in association
with Gecko Ltd, Bicester, Oxon

Printed in Italy

Introduction

This dictionary is written mainly for students in the 11–16 age group, but it should also be helpful to anyone seeking a basic knowledge of the vocabulary of mathematics, and there are many words from the fascinating byways of mathematics that are outside the strict limits of a school curriculum.

This is not a dictionary of etymology, grammar or English usage. So, for instance, there is no attempt to list every possible noun, adjective or verb, or all singular and plural forms—although there is a two-page spread devoted to word origins and plurals. The main purpose is to provide headwords in the form in which they are most often met in mathematics.

Any dictionary must use words to explain other words. There is no escape from this, and all users are assumed to have a grasp of non-mathematical English language. The real problem, which has been acknowledged since the time of Euclid, is that of defining the most basic words such as 'point', 'line', 'surface', and so on. These words are explained in this dictionary, but it has to be accepted that they are 'intuitive ideas' or 'common notions'. No matter where a start is made, understanding has to break in at some place.

Cross-referencing is always a problem in any dictionary which looks at words in isolation. For that reason this dictionary is divided into a number of themes, each contained on two facing pages. This helps readers to see how words relate to each other. For instance, it is much easier to read about the circle than to look at a series of separate entries, in different places, on words such 'circle', 'diameter', 'radius', 'sector', etc. In this way, each two-page opening of the book gives the reader a good account of a particular mathematical topic. There is a Wordfinder at the front, and that should be seen as the way in to the dictionary.

Compared with the first edition, this new edition has been enlarged by over two hundred new entries. These have been accommodated by the rewriting of many of the original themes and the addition of twenty new themes. Many extra formulas and symbols have also been provided. All this has led to an increase in the length of the wordfinder. Thus I have attempted to provide a wider, and in some cases deeper, coverage than was given by the first edition. It is hoped that this new edition will continue to provide a useful tool for the beginner while also giving support to the more advanced student.

Frank Tapson
March 1999

How to use this book

1. Look up the word in the Wordfinder at the front of the book. It looks like this and will give you the page number(s) you need in the Dictionary.

same word with different uses as shown by words in brackets

name of a section where several related words are to be found

Celsius scale 136, 170
centi- 153
centigrade *(angle)* 13
centigrade *(temperature)* 136
centre of rotation 142
centre of symmetry 130
chord 26, 49, 170
circle 26
circular cone 32
circumcircle 30
circumference 26, 110

page on which this word is to be found

appears on two pages although the meaning is the same – entries in order of significance

2. Look up the page number(s). All the entries are in two-page spreads organised by topic. Read the explanations which look like this:

word being explained

using a word which is explained elsewhere, often on the same page

rhombus A rhombus is a **quadrilateral** whose edges are all the same length; and usually no vertex *(=corner)* angle is a right angle. *Its diagonals bisect each other at right angles and both are also lines of symmetry.*

not needed for the explanation but to help with one word

words in italics are not part of the explanation but give further information

Some words or phrases have the same meaning:

power ≡ **index**

this is another word for this which has already been explained, usually on the same page

Sometimes one word can have two meanings which are slightly different, and it is shown like this:

to emphasise a particular word

circle A circle is EITHER a closed curve . . . .
OR it is the shape . . . .

Area of circle = $\pi \times$ Radius $\times$ Radius = πr^2

a formula is printed on a coloured background to make it stand out

3. Look at the diagrams (if there are any).

A

Wordfinder

Wordfinder

Wordfinder

Fermat   70
Fermat's last theorem   42
Fermat's problem   42
ferry problems   101
Fibonacci   70, 170
Fibonacci sequence   104
field   56
figurate numbers   88, 170
figure   14
finite   165
finite set   107
first quadrant   54
flow diagram   9
focus   34, 161, 170
foot   150
foot of a perpendicular   133
formula   7, 161, 168
formulas for shapes   48
four 4's   100
four-colour problem   47
fourth quadrant   54
fractions   50, 168
free vector   154
French curves   59
frequency   112, 94
frequency diagram   116
frieze patterns   84, 170
frustum of a cone   32, 161
frustum of a pyramid   96, 49
full turn   12
function   10
function machine   9
fundamental theorem of arithmetic   44
furlong   150

G

g   2
gallon   150
gamma (Γ, γ) 128
Gauss   71, 170
Gaussian integer   42
Gb   56
gelosia multiplication   22, 170
general angles   148
geodesic line   111, 170
geodetic surveying   124
geometric constructions   46, 170
geometric mean   163
geometric progression   104

geometric series   105
geometry   52, 168
giga-   153, 170
gill   150, 170
glide reflection   143
global maximum or minimum   24
GMT *or* G.M.T.   2
Gödel   71
Goldbach's conjecture   42
golden ratio   132
golden rectangle   98
googol *and* googolplex   79, 170
GP   2, 104
grade *or* grad   13
gradient   54
Graeco-Latin squares   101, 170
grain   150
gram   150
graphics calculator   57
graph *(in topology)*   140
graphs   54, 168
gravitational acceleration   153
great circle   74
Greek alphabet   128
Greek number system   81
Greenwich meridian   74
Gregory series   87
grid references   36
gross *(a quantity)*   79
gross *(in money or weight)*   17
group   123
grouped data   112

H

half-angle formulas *(for triangles)*   149
Hamiltonian walk   42
happy numbers   77
harmonic series   105
Harshad numbers   76
haversine   162
hcf *or* h.c.f.   2, 44
hectare   150
hecto-   153
hectometre   150
height   111
helix   38, 161
hemisphere   32
hendeca-   164
hept-   164

Wordfinder

Wordfinder

Wordfinder

Q

Q 82
QED 2
quad- 164
quadrants 54
quadratic equation 8
quadratic graph 55
quadrature 167
quadrilaterals 98, 90
quadrivium 167
qualitative data 114
quantitative data 114
quart 150
quartic 163
quartile 119
quin- 164
quincunx 167
quinquangle 167
quintic 163
quota sampling 115
quotient 21, 171
quotition 159, 171

R

ℝ 82
radial survey 125
radian 13
radius 26, 161, 171, 169
radius vector 36
RAM 56
Ramanujan's formula 32, 171
random 115
random sample 115
random selection 115
random sequence 104
random variations 114
range *(in algebra)* 10
range *(in statistics)* 118
rangefinder 60
rate of exchange 18
rate of interest 16
ratio 51, 169
rational numbers 82
raw data 112
real numbers 82
real variable 6
reciprocal 78, 171
reciprocal bearing 74

record 56
Recorde 70
recreational mathematics (1) 100
recreational mathematics (2) 102
rectangle 98, 48
rectangular coordinates 36
rectangular hyperbola 34
rectilinear shape 110
recurring decimal 50
recursive sequence 104
reduced fraction 51
reducible 9
reductio ad absurdum 65, 171
re-entrant polygon 90
reflection 142
reflective symmetry 131
reflex angle 12
region 140
regular polygon 90
regular polyhedron 92
regular quadrilateral 98
regular tilings 138
relative error 4
relative frequency 94
relatively prime 162
remainder 21
rep-tile 138
repunits 77, 171
residue 123, 171
resultant 154, 171
Retail Prices Index 19
retardation 62
Reuleaux polygon 111
reverse percentage 18
reverse Polish notation 167
rho (P, ρ) 128
rhomb 98
rhombohedron 108
rhomboid 98, 108
rhombus 98, 161, 171, 169
right angle 12, 169
right-angled triangle 144
right circular cone 32
right circular cylinder 32
right-handed system 157
right identity 123
right prism 96
right pyramid 96
ROM 56

Wordfinder

Wordfinder

The Dictionary

abbreviations and mnemonics

abbreviation An abbreviation is a shortened form of a word or phrase, often made by using the initial letter (or letters) of the word (or words). *Some of the more common abbreviations used in mathematics are given below.*

a.m. *(ante meridiem)* before noon or midday

AP arithmetic progression

APR annual percentage rate

cm centimetre(s)

cu cubic (referring to units)

dp or **d.p.** decimal places

GMT or **G.M.T** Greenwich Mean Time

GP geometric progression

g gram(s)

hcf or **h.c.f.** highest common factor

kg kilogram(s)

km kilometre(s)

L or **l** litre(s)

lcd or **l.c.d.** lowest (or least) common denominator

lcm or **l.c.m.** lowest (or least) common multiple

m metre(s)

mm millimetre(s)

mod modulus

p.m. *(post meridiem)* after noon or midday

QED *(quod erat demonstrandum)* which was to be proved

sf or **s.f.** or **sig. fig.** significant figures

SI Système International d'Unitès (international system of units)

sq square (referring to units)

UT or **U.T.** Universal Time. *Has replaced Greenwich Mean Time*

acronym An acronym is an **abbreviation** that is pronounceable and is usually said as a word. *An acronym is often written with capital letters, indicating that it is not a 'real' word.*

BODMAS is an **acronym** that serves as a reminder of the order in which certain operations have to be carried out when working with equations and formulas.

Brackets   Of   Division   Multiplication   Addition   Subtraction

SOHCAHTOA is an **acronym** that serves as a reminder of how the trigonometric ratios for a right-angled triangle are formed. The meaning is:

Sine A = Opposite ÷ Hypotenuse

Cosine A = Adjacent ÷ Hypotenuse

Tangent A = Opposite ÷ Adjacent

mnemonic   A mnemonic is a device which is intended to help a person's memory. *Some mnemonics are given below.*

trigonometric ratios   An aid to remembering how the trigonometric ratios for a right-angled triangle are formed (as in SOHCAHTOA) is the question:

Should Old Harry Catch Any Herrings Trawling Off America?

Euler's formula   A **mnemonic** to help in remembering Euler's formula for graphs giving the relationship between faces, vertices and edges is:

say	Fred	And	Vera	Took	Eric		Too
to *remember*	Faces	Add	Vertices	Take	Edges	=	Two

SI prefixes   A **mnemonic** to help remember the order of some SI prefixes is:

say	to	give me	kicks	my	musicians	now	play for	ages
to remember	tera	giga mega	kilo	milli	micro	nano	pico femto	atto

trigonometric values   The numerical values of the trigonometric ratios of certain angles are given by this easily remembered table:

$\theta =$	0°	30°	45°	60°	90°
$\sin \theta =$	$\dfrac{\sqrt{0}}{2}$	$\dfrac{\sqrt{1}}{2}$	$\dfrac{\sqrt{2}}{2}$	$\dfrac{\sqrt{3}}{2}$	$\dfrac{\sqrt{4}}{2}$

To get the cosine values, write the second line of values in reverse order.

π The first 8 digits of π are given by the number of letters in each of the words in the question below. Nine digits are given in the second sentence.

may	I	have	a	large	container	of	coffee ?	
3.	1	4	1	5	9	2	6	
see,	I	have	a	rhyme	assisting	my	feeble	brain
3.	1	4	1	5	9	2	6	5

$\dfrac{1}{\pi}$ The reciprocal of π to 6 decimal places is given by the number of letters in each of the words in this question:

can	I	remember	the	reciprocal ?
.3	1	8	3	10

e The first 8 digits of e are given by the number of letters in each of the words in this sentence:

to	express	e	remember	to	memorize	a	sentence
2.	7	1	8	2	8	1	8

$\sqrt{2}$ The first 4 digits of $\sqrt{2}$ are given by the number of letters in each of the words in the answer to the question:

the root of two?	I	wish	I	knew
1.	4	1	4	

division of fractions   An aid to remembering the process for dividing one fraction by another is the rhyme:

The fraction you are dividing by
Turn upside down and multiply   *Example:* $\dfrac{3}{4} \div \dfrac{9}{16} = \dfrac{3}{4} \times \dfrac{16}{9} = \dfrac{4}{3} = 1\dfrac{1}{3}$

coordinate pairs (x,y) A phrase to help remember the order for plotting the ordered pair (x,y) is 'along the passage and up the stairs', meaning 'go along to the x-value and up to the y-value'.

accuracy The accuracy of a number is an indication of how exact it is. *Very often in numerical work, and always when making measurements (unless counting distinct objects), the answer cannot be an exact one, so it is necessary to indicate just how accurate it is.*

approximation An approximation is a stated value of a number that is close to (but not equal to) the true value of that number. *Several reasonable approximations are always possible for any number. The one most suitable for the purpose in hand must be chosen.*
Example 3, 3.1 and 3.14 are all approximations to π (= 3.141 59...)

$\approx$ is the symbol for 'approximately equal to'   *Example: π ≈ 3.14*

estimation An estimation is an **approximation** of a quantity which has been decided by judgement rather than by carrying out the process needed to produce a more accurate answer. *The process might be measuring, doing a sum, or anything else of that nature.*
Examples: An estimation of the number of people in a room might be 30, when actual measurement (= counting) shows it is 27.
An estimation of the value of (23.7 × 19.1) ÷ 99.6 might be 4, or 5, or 4 and a bit. Doing the arithmetic suggests that 4.54 is a good approximation.

error The error is the difference between the value of an **approximation**, or an **estimation**, and the true value. *It may or may not have a plus (+) or minus (−) sign attached indicating whether it is too big or too small.*

absolute error The absolute error is the actual size of the **error** with NO sign.

relative error The relative error measures the size of the **error** as a fraction of the true value.
Example: π ≈ 3.14   *(True value is 3.141 592 653 5 ...)*
Absolute error is: True value − Approximate value = 0.001 592 653 5 ...
Relative error is: Absolute error ÷ True value
        = 0.001 592 653 5 ... ÷ 3.141 592 653 5 ... = 0.000 506 957...
Since π is itself a never-ending decimal, the dots are used to show that none of these values is an exact one. However, sufficient figures have been used to ensure the answer is accurate enough for all practical purposes.

percentage error The percentage error is the **relative error** as a percentage.
Example: In the above example, π = 3.14 gives an error of about 0.05%.

truncate To truncate something is to cut it short.
Example: In order to work with π in a calculation we always have to truncate the true value because it is a never-ending decimal.

truncation error A truncation error is an **error** introduced by **truncating** a number.

order of magnitude Two values are said to be of the SAME order of magnitude if their difference is small in relation to the size of the numbers being compared. *It is used rather loosely, and generally only with large numbers. Example: 32 million and 35 million have a difference of 3 million but, as this is less than 10% of either of them, it could be said that they are of the same order of magnitude.*

rounding is done when a number is **truncated** so as to minimise any error. *It is carried out on the last digit of the truncated number and is decided by the first digit of those being discarded. If the first of the discarded digits is 5 or more, then the last one of the truncated digits is increased by 1; otherwise no change is made.*
Example: π (= 3.141 59...) can be truncated to 3.14 or 3.142, etc.

rounding error A rounding error is an **error** introduced by **rounding** a number.

to the nearest '...' indicates that an **approximation** has been made by **rounding** as necessary, so that the given value finishes on a digit whose **place-value** is stated in '...'. *Usually this is done only with whole numbers, but it can be applied to decimal fractions (to the nearest tenth etc.). Rounding must be done.*
Example: The true attendance at a football match was 24 682, but such a number might be given as 24 680 (to the nearest 10) or 24 700 (to the nearest 100) or 25 000 (to the nearest 1000).

to '...' decimal places (dp) indicates an **approximation** has been made by **truncation** to leave only the number of digits after the decimal point stated in '...' *Rounding is usually done. It is correct to use the = sign rather than ≈*
Example: π = 3.14 (to 2 dp)
π = 3.141 59 (to 5 dp)

significant figures (sf) are used to express the relative importance of the digits in a number; the most important being the first digit, starting from the left-hand end of the number, which is not zero. *Starting with the first non-zero digit all digits are then counted as significant up to the last non-zero digit. After that, zeros may or may not be significant; it depends on the context.*
Examples of:

4 significant figures	*1234*	*78 510*	*16.32*	*0.024 71*	*0.005 026*
3 significant figures	*1230*	*78 500*	*16.3*	*0.024 7*	*0.005 03*
2 significant figures	*1200*	*79 000*	*16*	*0.025*	*0.005 0*
1 significant figure	*1000*	*80 000*	*20*	*0.02**	*0.005*

*Note that the number 0.02 has been rounded according to the evidence of the value at the top of its column and not the one immediately above it.

to '...' significant figures indicates an **approximation** has been made by **truncation** to leave only the number of significant figures stated in '...'. *Rounding should be done.*
Examples: π = 3.14 (to 3 s.f.) π = 3.14159 (to 6 sf)

nominal value The nominal value of a number is the 'named' amount or, in the case of a measurement, the size it is intended to be.

tolerance is the amount by which a **nominal value** may vary. *The greatest use of this is in the manufacturing industries.*
Example: A piston diameter is given as 65 mm ± 0.015 mm, so it has a nominal diameter of 65 mm but could be as small as 64.985 mm or as big as 65.015 mm.
The error allowed is not always evenly spread, as in $76^{+0.023}_{-0.037}$ mm.

algebra is the branch of mathematics that deals with generalised arithmetic by using letters or symbols to represent numbers. *Any statement made in algebra is true for ALL numbers and not just specific cases.*
Example: $9^2 - 6^2 = 81 - 36 = 45$ is true for those numbers.
But, $x^2 - y^2 = (x + y)(x - y)$ is true for all numbers.
and $9^2 - 6^2 = (9 + 6)(9 - 6) = 45$ is just one example of many.

convention for letters In **algebra** the convention is that letters for **variables** are taken from the end of the alphabet and are usually lower case (x, y, z), while those representing **constants** are taken from the beginning of the alphabet (A, B, C, a, b, c). *This convention is not followed for formulas where the letters used are those which best serve as reminders of the quantities being handled.*

variable A variable is a symbol (usually a letter such as x, y, z) that may take any value from a given range of values. *Unless the range of possible values is stated, then the variable can be any real number.*

real variable A real variable is a **variable** whose values must be **real numbers**.

constant A constant is a value that is unchanged whenever it is used for the particular purpose for which it was defined. *Usually it is given as a number, but in some cases a letter might be used to indicate that a constant (of the correct value) must be put in that place.*

coefficient A coefficient is a **constant** attached in front of a **variable**, or a group of variables, where it is understood that once the value of the variable(s) has been worked out, then the result is to be multiplied by the coefficient. *The absence of a coefficient is equivalent to a 1 being present.*
Example: In $3x$ $7xy$ Ax^2y y^2 the coefficients are 3, 7, A and 1

expression An expression in **algebra** is most often a collection of quantities, made up of **constants** and **variables**, linked by signs for operations and usually not including an equals sign. *In practice it is an imprecise word and is used very loosely.*
Examples: $x + y$ $3 + x^2 - y$ $4(x - y)$ $3x^2 + 5y$ are all expressions.

literal expression A literal expression is an **expression** in which the **constants** are represented by letters as well as the variables.
Examples: $Ax^2 + Bx + C$ and $ax + b$ are literal expressions.

term The terms in a simple algebraic **expression** are the quantities that are linked to each other by means of + or − signs. *In more complicated expressions the word 'term' is given a much looser meaning.*
Example: $5x^2 + 3x - xy + 7$ has four terms.

like terms are those **terms** that are completely identical in respect of their **variables**. *They must contain exactly the same variables and each variable must be raised to the same power. The coefficients can be different. The purpose of identifying like terms is so that they may be collected together by addition or subtraction.*
Examples: Some pairs of like terms are $3x$ and $5x$; $7x^2y$ and x^2y

constant term A constant term in an **expression** or **equation** is any **term**, or terms, consisting only of a **constant** with no **variables** attached. *Often it is merely referred to as a 'constant' of the equation.*
Example: In $5x + y - 2$ the coefficients are 5 and 1; the constant term is $^-2$

equation An equation is a statement that two **expressions** (one of which may be a constant) have the same value.
Examples: $2x + 7 = 15$ and $3(x + 5) = 3x + 15$ are both equations.

conditional equation A conditional equation is an **equation** which is only true for a particular value, or a number of values, of the variable(s).
Example: $2x + 7 = 15$ is only true when $x = 4$

identity An identity in **algebra** is an **equation** that is true for ALL values of the variable(s). *Strictly speaking, the $\equiv$ sign should be used instead of the $=$ sign, to show that it is an identity, but usually this is only done for emphasis.*
Examples: $x^2 - y^2 \equiv (x + y)(x - y)$ $\qquad 3(x + 5) \equiv 3x + 15$

formula A formula is a statement, usually written as an **equation**, giving the exact relationship between certain quantities, so that, when one or more values are known, the value of one particular quantity can be found.
Example: For a sphere of diameter d the volume V can be found from the formula $V = \pi d^3 \div 6$

transpose To transpose an **equation** or **formula** is to rearrange it (under definite rules) to produce an equivalent version. *This is usually done in order to simplify it or make it easier to work with.*
Example: The equation $4x + y = 5$ can be transposed to $y = 5 - 4x$

changing the subject of a formula is **transposing** it so that the value of a different quantity from that given can be worked out.
Example: The formula $A = \pi r^2$ is transposed to give $r = \sqrt{\dfrac{A}{\pi}}$

simplify To simplify an algebraic **expression**, gather all **like terms** together into a single term.
Example: $2x^2 + 9 - 7x + 3xy + 5x + x^2 - 1$ simplifies to $3x^2 - 2x + 3xy + 8$

substitution A substitution in algebra is done by replacing one **expression**, or part of an expression, by something of equivalent value so that the overall truth of the original expression is unchanged.
Example: Given $4y + 3x = 22$ and $y = 2x$, then the 2nd expression can be substituted for y in the 1st to give $4(2x) + 3x = 22$, simplified to $11x = 22$

elimination of a **variable** from an **expression** is the removal of that variable and is usually done by **substitution**.
Example: In the previous example under 'substitution' the y in the 1st expression was eliminated by the use of $y = 2x$

nested multiplication is a way of rewriting an **expression** so that it is easier to work with when calculating values. *This is especially useful with a calculator where the value of the variable can be kept, and recalled from, the memory.*
Example: $4x^3 - 5x^2 + 7x - 8$ is easier to use as $[(4x - 5)x + 7]x - 8$

degree of a term   The degree of an algebraic **term** is found by adding together ALL the powers of the **variables** in that term.
Examples: $2x^3$ has degree 3; $4x^3y^2$ has degree 5; $3xy$ has degree 2

degree of an expression   The degree of an **expression** is given by the highest value found among the degrees of all the terms in that expression.
Example: $x^4 - 4x^3y^2 + 6y^2$ is an expression of degree 5 (the middle term).

linear equation   A linear equation is an **equation** involving only an **expression**, or expressions, of **degree** 1. *Such an equation can be represented graphically by a straight line.*
Examples: $y = 3x + 2$ $y = 4$ $x = 3y - 5$ are all linear equations.

quadratic equation   A quadratic equation is an **equation** involving an **expression**, or expressions, containing a single variable, of **degree** 2.
Examples: $x^2 + 3x - 5 = 0$ $3(x + 1)^2 = 0$ $4x^2 - 3x + 4 = 0$

satisfy   When a value is substituted for a variable in an equation and leaves the truth of the equation unchanged, that value is said to satisfy that equation.
Example: $x = 2$ satisfies the equation $3x + 7 = 13$ since $(3 \times 2) + 7 = 13$

solution   The solution(s) of an **equation**, or a set of equations, is (are) the value, or values, of the variable(s) that will **satisfy** the equation(s).
Example: $4x - 5 = 7$ has the solution $x = 3$ since $(4 \times 3) - 5 = 7$

unique solution   A unique solution is the ONLY **solution**.
Example: $2x + 5 = 13$ has the unique solution $x = 4$
$x^2 = 9$ has one solution $x = 3$ but it is not unique since $x = {}^-3$ is another.

trivial solution   A trivial solution is one which is obvious and of little interest.
Example: $x^n + y^n = z^n$ has a trivial solution $x = y = z = 0$. It has others.

root   A root of an **equation** is a value that will **satisfy** the equation which has been formed by putting an **expression**, containing one variable, equal to zero. *The maximum number of roots possible is the same as the **degree** of the expression. Roots may be **real** or **complex** numbers.*
Examples: $x^2 - 8x + 15 = 0$ has the 2 roots: $x = 3$ or 5
$x^3 - 4x^2 - x + 4 = 0$ has the 3 roots: $x = {}^-1,\ 1$ or 4

trial and improvement   is a method of looking for a **solution** in which a guessed-at value is put into a problem; the consequences are followed through, and, on the basis of any error found, a better guess-value is made. *It is a very powerful method and capable of solving almost any kind of problem, only provided that it is solvable. It is also known as 'trial and error'.*
Example: To find a root of $2x^3 + 4x - 5 = 0$ first try $x = 1$ in the expression. This gives a value of 1. Trying $x = 1.1$ gives a value of 2.062 so try $x = 0.9$ to get 0.058. Perhaps this is close enough to 0?

independent equations   A set of equations is independent if no single **equation** in the set can be made from some combination of the others.
Example: $x + y = 7$; $2x - z = 6$ and $3x + y - z = 13$ are not independent since the 3rd can be made by adding the first two equations. Changing the 3rd to $3x + y + z = 13$ would make an independent set.

simultaneous equations A set of simultaneous equations consists of two (or more) **equations** whose **variables** all take the same value at the same time. *Provided all the equations are* **independent**, *and a solution is possible, then n simultaneous equations containing n variables will have a unique solution. There are several ways of solving simultaneous equations, the most common being by a combination of substitution and elimination.*

indeterminate equation(s) An indeterminate equation is an **equation** (or a set of equations) for which any number of solutions can be found.
Example: $x + 2y = 3$ has solutions: (x,y) (1,1) (2,0.5) (3,0) (4, ⁻0.5) ...

Diophantine equations are **indeterminate equations** having only whole numbers for coefficients and having only whole numbers as acceptable solutions. *If the previous example were a Diophantine equation, it could only have solutions such as 1,1;   3,0;   5, ⁻1;   ...*

factors The factors of an **expression** in algebra are two, or more, other expressions which can be multiplied together to produce the original expression.
Example: $(x + 1)$ and $(x - 2)$ are factors of $x^2 - x - 2$ since
$$(x + 1)(x - 2) \equiv x^2 - x - 2$$

reducible A reducible **expression** is one which has at least two **factors**.
Example: $x^3 - 4x^2 + 3x - 12$ can be reduced to $(x^2 + 3)(x + 4)$

irreducible An irreducible **expression** is one that cannot be **reduced**.

expansion The expansion of an **expression** is carried out by doing as much as possible to make it into a collection of terms connected only by + and − signs. *Usually this entails doing as much multiplication as can be done, and removing all the brackets.*
Example: $(3x + 1)(x - 2) + 4(x - 5)$ expands to $3x^2 + x - 6x - 2 + 4x - 20$

multinomial A multinomial expression is one having two or more terms.

binomial A binomial expression is a **multinomial** having two terms.
Examples: $3x + 4$ $x - y$ $5 - 7y$ are all binomial expressions.

trinomial A trinomial expression is a **multinomial** having three terms.
Examples: $3x^2 - 5x + 4$ $x - y + 7$ are trinomial expressions.

polynomial expression A polynomial expression is an **expression** made of two (or more) **terms** where each term consists of a **coefficient** and a **variable** (or variables) raised to some non-negative power which must be a whole number. *The non-negative powers could be zero.*
Examples: $4x^3 - 5xy^2 + y^3 + 2$ is a polynomial but $x^2 + x^{-1}$ is not.

flow diagram A flow diagram is a drawing intended to make clear the order in which operations have to be done so as to produce a result. *While they can be used to explain any production process, they are most commonly used in mathematics as an aid to working out values of functions, formulas, etc.*
Example: $y = 3x^2 + 5$ is shown in this flow diagram:

Enter x → Square it → × 3 → + 5 → gives y

function machine ≡ **flow diagram**. *Usually drawn in an informal style.*

algebra (functions)

mapping A mapping is the matching of the **elements** from one **set** to the elements of another set by use of a rule. *The elements are usally numbers of some type (integer, real, complex, etc.) or they could be algebraic.*

mapping diagram A mapping diagram is a drawing used to show the effect of a **mapping** by listing the two sets and drawing arrows indicating how the elements are to be matched.
Example: The mapping diagram (right) shows what happens for the rule 'multiply by 3 and add 1' for some values.

$$1 \rightarrow 4$$
$$2 \rightarrow 7$$
$$3 \rightarrow 10$$
$$4 \rightarrow 13$$

one-to-one correspondence A one-to-one correspondence occurs when a **mapping** between two sets of the SAME size pairs all the elements of each set without using any element twice. *The mapping diagram above shows a one-to-one correspondence.*

domain The domain is the set that the **mapping** is coming FROM.

codomain The codomain is the set that the **mapping** is going TO.

range The range of a mapping is made up of those elements in the codomain which are actually used in the mapping.
Example: The mapping on the right uses the same set for both the domain and codomain and the rule 'multiply by 2'. The range is only the even numbers.

many-to-one correspondence A many-to-one correspondence occurs when a **mapping** matches more than one element in the **domain** with the same element in the **codomain**.
Examples: The rule 'is the number whose square is' would match both 2 and ⁻2 from the domain with 4 in the codomain. The rule 'is the child whose mother is' could match more than one child to one woman.

one-to-many correspondence A one-to-many correspondence occurs when a **mapping** matches one element in the **domain** with more than one element in the **codomain**.
Examples: The rule 'is the number whose square root is' would match 4 from the domain with both 2 and ⁻2 from the codomain.
The rule 'is the mother of' could match one woman to more than one child.

function A function is a **mapping** which involves either a **one-to-one correspondence** or a **many-to-one correspondence**. *The sets to be used for the domain and codomain must be defined; they can be the same.*
Example: For positive numbers the mapping 'is the square of' is a function.

f(x) is the symbol for a **function** involving a single **variable** identified in this case as **x**. *This is only a general statement and a definition of what f(x) actually does is needed before it can be used. It is usually defined by means of an algebraic expression.*
Examples: $f(x) \equiv 3x + 1$ $f(x) \equiv 4x^2 + 3x - 7$ $f(x) \equiv 2x(x - 8)$

$y = f(x)$ and $f : x \rightarrow y$ are two ways of saying that there is a **function** of x which produces a **mapping** from x-numbers to y-numbers, though neither says how the mapping is actually done.
Example: Given $f(x) \equiv 2x + 3$, then $y = f(x)$ is the same as $y = 2x + 3$ or $f{:}x \rightarrow 2x + 3$ and both produce the same mapping (or table of values).

inverse function An inverse function is a second **function** that reverses the direction of the **mapping** produced by a first function. *For an inverse to exist, the first function must produce a one-to-one correspondence. Sometimes a function can be forced into being of the one-to-one type by restricting the numbers to be used in the domain and codomain.*
Example: $f(x) \equiv x^2$ is a function of x producing a many-to-one correspondence since x or ^-x will both produce the same value of $f(x)$. This means that, though the mapping can be reversed (using square roots), the reverse mapping is one-to-many and therefore not a function. If the restriction is made that only positive numbers are allowed, then $f(x)$ is a one-to-one correspondence and an inverse function exists.

$f^{-1}(x)$ is the symbol for the **inverse function** of $f(x)$.

independent variable The independent variable in a **mapping** is the element or number FROM which the mapping STARTS. *In the mapping (right) for $f{:}x \rightarrow x^2 - 1$ the values of the independent variable are {1, 3, 4, 5, 6, 7, 8}. The usual symbol is x.*

$$\begin{array}{ccc} 1 & \rightarrow & 0 \\ 3 & \rightarrow & 8 \\ 4 & \rightarrow & 15 \\ 5 & \rightarrow & 24 \\ 6 & \rightarrow & 35 \\ 7 & \rightarrow & 48 \\ 8 & \rightarrow & 63 \end{array}$$

dependent variable The dependent variable in a **mapping** is the element or number TO which the mapping GOES. *In the mapping (right) for $f{:}x \rightarrow x^2 - 1$ the values of the dependent variable are {0, 8, 15, 24, 35, 48, 63}. The usual symbol is y or $f(x)$.*

explicit function An explicit function is a **function** that is given entirely in terms of the **independent variable**.
Examples: $f(x) \equiv 2x + 5$ $y = x^2 + 3$ $f{:}x \rightarrow x(x-1)$ are all explicit.

implicit function An implicit function is a **function** that is given in terms of both the **independent** and the **dependent variables**. *Implicit functions are usually written in a way that equates them to zero.*
Example: $x^2 + 2xy - y^2 = 0$

$F(x,y)$ is the symbol for an **implicit function** involving two variables, identified in this case as x and y.

bounds are two limits that values of a particular function cannot be greater or less than. The UPPER bound is a limit above which the function can produce no higher values. The LOWER bound is a limit on the lowest values. *It is usual to make the bounds as tight as possible. It may, or it may not, be possible for values to actually touch the stated bounds.*
Example: Given $f(x) \equiv 2(\sin x + 1)$, then $f(x)$ has an upper bound of 4 and a lower bound of 0 since its value cannot go above or below those limits.

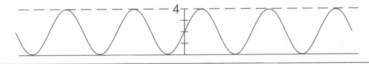

angles

angle An angle is made when two straight lines cross or meet each other at a point, and its size is measured by the amount one line has been turned in relation to the other.

full turn A full turn is a measure of the **angle** made when the line which is turning has moved right around and returned to its starting position.

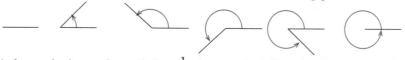

degree A degree is the angle made by $\frac{1}{360}$ th part of a **full turn**. *This means that there are 360 degrees in a full turn or, as it is usually said, 'There are 360 degrees in a circle'.*

minute A minute is the angle made by $\frac{1}{60}$ th part of a **degree**.

second A second is the angle made by $\frac{1}{60}$ th part of a **minute**.

There is an increasing tendency to use degrees and decimal fractions rather than minutes and seconds, except in navigation.

 °   ′   ″   are the symbols for degree, minute and second respectively.
Example: 103° 26′ 47″ is 103 degrees 26 minutes 47 seconds.
In decimal form this is about 103.446°.

right angle A right angle is the angle made by one-quarter of a **full turn** or 90°. *It is usually shown on drawings by means of a small square in the corner.*

straight angle A straight angle is the angle made by one-half of a **full turn** and is equal to 180°. *It looks exactly the same as a straight line.*

acute angle An acute angle is one which is LESS than a **right angle**.

obtuse angle An obtuse angle is one which is MORE than a **right angle** but LESS than a **straight angle**.

reflex angle A reflex angle is one which is MORE than a **straight angle** but LESS than a **full turn**.

complementary angles are a pair of angles which add together to make 90°.
Example: Angles of 30° and 60° are complementary.

complement The complement of an acute angle is the amount needed to be added on to make 90°.   *Example: The complement of 70° is 20°.*

supplementary angles are a pair of angles which add together to make 180°.
Example: Angles of 30° and 150° are supplementary.

supplement The supplement of an angle is the amount needed to be added on to make 180°.   *Example: The supplement of 70° is 110°.*

conjugate angles are a pair of angles which add together to make 360°.

grade or **grad** A grade is the angle made by $\frac{1}{100}$ th part of a **right angle**.

centigrade A centigrade is the angle made by $\frac{1}{100}$ th part of a **grade**.

> *Grades and centigrades are now little used for angle measurement, though they are often found on calculators under the label* GRA.

radian A radian is the angle made at the centre of a circle between two radii when the length of the arc on the circumference between them is equal to the length of one radius. *This unit of angle measurement is used a lot in further mathematical work.*

> There are 2π radians in a full turn.
> There are π radians in $180°$
> To change radians into degrees multiply by 180 and divide by π
> To change degrees into radians multiply by π and divide by 180

positive angle A positive angle is one measured in the anticlockwise direction.

negative angle A negative angle is one measured in the clockwise direction.

> *Whether an angle is positive or negative makes no difference to its actual size, which is the same in either direction, but is sometimes needed when a movement is being explained.*

angle of elevation The angle of elevation of an object is the angle through which someone must look UP from the **horizontal plane** to see that object.
> *Example: For a person standing on level ground who sees the top of a mast at an angle of elevation of 25° the situation is represented in this diagram:*

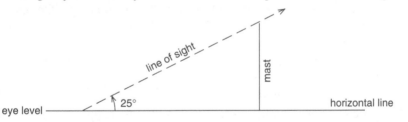

angle of depression The angle of depression of an object is the angle through which someone must look DOWN from the **horizontal plane** to see that object.
> *Example: For a person standing on a cliff top looking at a small boat and seeing it at an angle of depression of 25° the situation is represented in this diagram:*

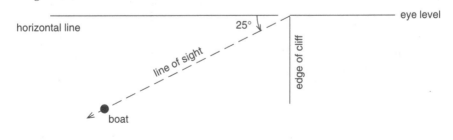

arithmetic (basics)

arithmetic is a part of mathematics that deals with the properties and handling of numbers, and their use in counting and calculating.

numbers are the basic elements of **arithmetic** which are used for expressing and recording, quantities or measures of various kinds.
Examples: 2 people, 7.8 metres, mark the point ($^-3$, 4)

digit The digits are the single symbols 0, 1, 2, 3, 4, 5, 6, 7, 8, 9 as used in everyday **arithmetic**. *Digits are also numbers but, more importantly, they are put together to make numbers.*
Example: 2167 is a number made of four digits.

numeral ≡ **digit**.

figure A figure may be either a **digit** OR a **number**. *Both 6 and 76 are figures.*

whole number A whole number is a **number** which has no fraction attached.
Examples: 8, 13, 207 are whole numbers.   2.5 is not a whole number.

consecutive numbers are **whole numbers** that follow each other in order when arranged in a sequence from smallest to largest.
Examples: 3, 4, 5 and   19, 20, 21, 22   are both groups of consecutive numbers.

even numbers are **whole numbers** which, when divided by 2 have no remainder. *Any number which ends in 0, 2, 4, 6 or 8 must be even.*
Examples: 20,   348,   1356   are all even numbers.

odd numbers are **whole numbers** which, when divided by 2 have a remainder of 1. *Any number which ends in 1, 3, 5, 7 or 9 must be odd.*
Examples: 17,   243,   8645   are all odd numbers.

parity The parity of a **number** refers to the fact of it being either **even** or **odd**.
Examples: The numbers 4 and 10 have the same parity (both even); and so also do 3, 7, and 15 (all odd); while 5 and 8 are of opposite parity.

square To square a number is to multiply it by itself. *The square of 1 is 1*
Examples: To square 6 work out 6 × 6 = 36
To square 2.5 work out 2.5 × 2.5 = 6.25

x^2 ^{2} is the symbol meaning '**square** the number that appears in the place of x'
Example: When $x = 1.3$, then x^2 means $1.3 \times 1.3 = 1.69$

square root A square root of a number is another number which when **squared** will equal the first number. *The square roots of 1 are 1 and $^-1$.*
Example: One square root of 16 is 4 since 4 × 4 = 16; another is $^-4$

$\sqrt{}$ is the symbol meaning 'the **square root** of the number given'.
Examples: $\sqrt{49} = 7$ (7 × 7 = 49) $\sqrt{3.24} = 1.8$ (1.8 × 1.8 = 3.24)

surd A surd is the square root of a whole number which produces an **irrational number**. *It can also be a cube (or other) root and is sometimes applied to an expression which contains a surd or surds. The square root of any prime number is a surd.*
Examples: $\sqrt{2}$ ($\approx 1.414 ..$) $2\sqrt{3}$ $\sqrt{19}$ are all surds.

perfect square A perfect square is a number whose **square root** is a **whole number**.
  Examples: 1, 4, 9, 16, 25, 36 and 289 are all perfect squares (with square roots of 1, 2, 3, 4, 5, 6 and 17 respectively).

cube To cube a number is to multiply it by itself and then multiply the result of that by the original number. *The cube of 1 is 1.*
  Examples: To cube 4 work out $4 \times 4 \times 4 = 64$
                To cube $^-4$ work out $^-4 \times {}^-4 \times {}^-4 = {}^-64$

x^3 ^{3} is the symbol meaning '**cube** the number that appears in the place of x'
  Example: When $x = 1.7$ then x^3 means $1.7 \times 1.7 \times 1.7 = 4.913$

cube root The cube root of a number is another number which, when **cubed**, will equal the first number. *The cube root of 1 is 1*
  Example: The cube root of 8 is 2 since $2 \times 2 \times 2 = 8$

$^3\sqrt{}$ is the symbol meaning 'find the **cube root** of the number given'.
  Examples: $^3\sqrt{8} = 2$ $^3\sqrt{4.096} = 1.6$ since $1.6 \times 1.6 \times 1.6 = 4.096$

digit sum The digit sum of a number is found by adding all its **digits** together.
  Example: The digit sum of 742 is $7 + 4 + 2 = 13$.

digital root The digital root of a **positive whole number** is made by finding its **digit sum** to make a new number and repeating this process on each new number made until only a single digit remains – this is the digital root of the original number.
  Example: $8579 \rightarrow 8 + 5 + 7 + 9 = 29 \rightarrow 2 + 9 = 11 \rightarrow 1 + 1 = 2$
                    so 2 is the digital root of 8579

casting out 9's is a method of checking on the accuracy of some arithmetic processes. *For the processes of addition, subtraction or multiplication of numbers, if the same process is applied to the digital roots of those numbers, then the digital root of that answer should be the same as the digital root of the actual answer obtained from the full numbers. If there is an error, this method will not identify where it is, only that there is one.*
  Examples: In these sums $_{(x)}$ is the digital root of the preceding number.

  $806_{(5)} + 57_{(3)} = 863_{(8)}$ $[5 + 3 = 8]$ ✓
  $806_{(5)} - 57_{(3)} = 749_{(2)}$ $[5 - 3 = 2]$ ✓
  $806_{(5)} \times 57_{(3)} = 45942_{(6)}$ $[5 \times 3 = 15 \rightarrow 1 + 5 = 6]$ ✓

order of operations From the expression $2 + 4 \times 3 - 1$ it is possible to get different answers according to the order in which the operations are done. Working from left to right $(+ \times -)$ gives 17; using $(+ - \times)$ gives 12; using $(\times + -)$ gives 13; and $(- \times +)$ gives 10. To prevent this happening there is an established order in which operations MUST be done.
  Anything in brackets has to be done first, then division and multiplication, and then addition and subtraction. An aid to remembering this is **BODMAS**.

  Examples: $(2 + 4) \times 3 - 1 = 17$ $(2 + 4) \times (3 - 1) = 12$
              $2 + 4 \times (3 - 1) = 10$ $2 + 4 \times 3 - 1 = 13$
  Even when not strictly necessary, brackets can be helpful.

arithmetic (commercial)

per cent A value given per cent means that the number stated is to be used to make a fraction with that number on the top and 100 on the bottom.
Example: 45 per cent is the fraction $\frac{45}{100}$

% is the symbol for per cent. *Example: 37% is 37 per cent or $\frac{37}{100}$*

per mil A value given per mil means that the number stated is to be used to make a fraction with that number on the top and 1000 on the bottom.
Example: 68 per mil is the fraction $\frac{68}{1000}$

‰ is the symbol for per mil. *Example: 21‰ is 21 per mil or $\frac{21}{1000}$*

percentage point A percentage point is the actual value of the the difference between two percentages, and NOT the percentage change.
Example: A report that 'The government reduced the interest rate of 5% by 1 percentage point' means that the new rate is (5 – 1)% = 4% and not 5% reduced by 1% (of 5%) to make it 4.95%.

principal The principal is the amount of money involved (usually at the start) in some transaction such as lending, borrowing or saving.
Example: On opening an account with £250 there is a principal of £250

interest The interest is the amount of extra money paid in return for having the use of someone else's money.

rate of interest The rate of interest states how the **interest** is to be worked out. *It is usually stated as a percentage of the principal for each given period. Example: The interest on the loan is set at 2% per month.*

simple interest In calculating simple interest at the end of each **period** it is always worked out only on the **principal** of the original amount. *This method is very little used nowadays.*
Example: A loan of $300 at 2% per month at simple interest would mean that at the end of each month interest of $6 would be owing. So, at the end of a year, the total interest owing would be $72 (plus the loan which still has to be repaid).

$$\text{Total simple interest} = \frac{\text{Principal} \times \text{Rate of interest (\%)} \times \text{Number of periods}}{100}$$

compound interest In calculating compound interest at the end of each **period** it is worked out on the **principal** plus ANY PREVIOUS INTEREST already earned. *This is the usual method nowadays. In this case, it is the total amount owing that is calculated, rather than the separate interest.*

$$\text{Total amount owing} = \text{Principal} \times \left(1 + \frac{\text{Rate of interest (\%)}}{100}\right)^{\text{Number of periods}}$$

depreciation The depreciation of the value of an object is the amount by which that value has fallen. *It is similar to compound interest, except that the value is decreasing. If the rate of depreciation (as a % per period) is known, then the new value can be worked out from this formula:*

$$\text{New value} = \text{Original value} \times \left(1 - \frac{\text{Rate of depreciation (\%)}}{100}\right)^{\text{Number of periods}}$$

discount A discount is an amount which is taken OFF the price of something. *Discounts are usually stated as percentages.*
Example: If a price of £12 has a discount of 10%, £1.20 is taken off.

cash discount A cash discount is a **discount** which is sometimes given if the amount owing is paid in cash, usually within a certain time.

gross A gross amount (of weight or money) is that total which exists at the beginning BEFORE any deductions are made for any reason.
Example: A person's wages are £270 (gross) but various taxes totalling £75 are taken off, so £195 is left. (£195 is the **net** *wage.)*

net A net amount (of weight or money) is the amount remaining AFTER any necessary deductions have been made.
Example: A packet containing rice weighs 760 grams (gross) but the packet weighs 40 grams. The net weight (= weight of the contents) is 720 grams.

overheads are those costs involved in running a business which are not directly connected with the products being handled by that business. *Items such as rent, rates, heating, lighting, administration, research and many others would be classed as overheads, though it is not always easy to know where the line should be drawn.*

selling price The selling price of an article is the price at which that article is offered for sale.

cost price The cost price of an article is the cost of making or purchasing that article. *If the article has simply been bought (and not made), then it might be called the* **buying price**. *Notice that in the process of buying and selling, say from A to B, what is the selling price for A is the cost price for B.*

profit is a measure of the gain made in a transaction or enterprise. *The measure is usually a financial one. In the commercial world, working out the actual profit is usually complicated by the difficulty of assigning the* **overheads**. *In the simple case of a single article where both the* **selling** *and* **cost price** *are known, the profit is the difference between the two, and is usually expressed as a percentage of the cost price.*

$$\text{Percentage profit} = \frac{\text{Selling price} - \text{Cost price}}{\text{Cost price}} \times \frac{100}{1}$$

The profit could be a negative amount, in which case it is more generally known as a **loss**.

mark-up is a term used, mainly in the retail trade, to indicate what has to be added on to the **cost price** of an article to find the **selling price**. *It is usually expressed as a percentage of the cost price.*
Example: A shopkeeper buys T-shirts at £2.50 each and uses a mark-up of 150% (= £3.75) to fix the selling price at £6.25 (£2.50 + £3.75).

pro rata is a Latin term used to mean that something has to be changed or shared 'in proportion'.
Example: Two people buy something for $10, one paying $7 and the other $3. They sell it for $20 and share the money 'pro rata' according to what each paid, which gives them $14 and $6 respectively.

reverse percentage is the name of the technique used when finding the original value of something, knowing only its current value and the percentage (of the original value) by which its original value was INCREASED. *This is a common problem when a price including a tax is known and the price without the tax is needed.*

> Original value = [100 × Current value] ÷ [100 + R]
> where R is the percentage change made to the original value.
> The second bracket is [100 − R] if a DECREASE had been made.

Example: A lawnmower costs £376 which includes VAT at 17.5%. Its cost without VAT must be [100 × 376] ÷ [100 + 17.5] = £320.

instalment An instalment is an amount of money paid at regular intervals over some agreed period of time. *Instalments commonly arise in connection with repaying a loan (plus interest) or keeping up a hire purchase agreement. It is usual to arrange instalments of equal size (payable weekly or monthly) so that the loan and the interest are paid off together. The size of instalment needed to do this can be calculated from this formula:*

> $$\text{Instalment} = \frac{A\, r\, F^n}{F^n - 1} \quad \text{where}$$
>
> A = Amount of loan
> r = Rate of interest (%) per period ÷ 100
> $F = r + 1$
> n = number of periods

Example: £300 loaned for 12 months at 2% per month compound interest.
$A = 300, \quad r = 0.02, \quad F = 1.02, \quad n = 12$
Monthly instalments = £28.37 (to nearest p)

APR The APR is the **Annual Percentage Rate** of a loan. *Organisations which lend money express their offers in various ways in order to make them look different and attractive. The law requires that the APR is also given so that comparisons can be made more easily. There are formulas for this.*

rate of exchange The rate of exchange between two systems is a statement of how a value in one system may be given as an equivalent value in the other system. *It is most often used in changing money between countries. Example: The rate of exchange between French (francs) and British (pounds) currency might be given as 8.26 francs to the pound. Then 47 pounds would be worth 47 × 8.26 = 388.22 francs and 150 francs would be worth 150 ÷ 8.26 = 18.16 pounds.*

bureau de change A bureau de change is a place where currencies of different countries can be exchanged. *Most banks have a bureau de change on their premises as do many travel agents, but others exist by themselves.*

income tax is a tax which is taken from a person's earnings by the government of the country to help them pay for all the things needed to run the country. *It is not the only way the government gets its money – there are other forms of tax (**VAT**, inheritance tax, capital gains tax, national insurance etc.).*

VAT or **Value Added Tax** is a tax paid on the goods or services bought by a customer, which is then paid by the supplier of those goods or services to the government. *The rate at which VAT is charged is given as a percentage, and varies with different goods and in different countries.*

RPI or **Retail Prices Index** The RPI is a measure of how the cost of things that people buy to support their everyday life is changing. *The figure is produced by noting the overall change in the cost of a large sample of items. It is not only food that is looked at, but also lighting, heating, travel, entertainment etc; in short, nearly everything that the ordinary person spends his or her money on. The whole sample is carefully balanced to reflect the importance that each item might be expected to have in the average person's life. The base value is set at 100 in one particular year and the changes are measured against that.*

Every country in the world carries out such a survey at least once a year. In the UK the survey (started in 1914) is done every month by the Office for National Statistics based on a total of over 100 000 items.

The table gives the RPI (in January) for the UK for each year since 1987, which was a base year. From this it can be seen that the cost from 1987 to 1992 has gone from 100 to 135.6 or, more usefully, the cost in 1992 is 1.356 times as much as in 1987. It is also known as the **Consumer Price Index** or the **Cost of Living Index**

1987	100
1988	103.3
1989	111.0
1990	119.5
1991	130.2
1992	135.6
1993	137.9
1994	141.3
1995	146.0
1996	150.2
1997	154.4
1998	159.5
1999	163.4
2000	166.6

inflation is a measure of the rate at which the cost of goods is changing over some period of time and is usually expressed as a percentage.

$$\text{Rate of Inflation (\%)} = \frac{E - B}{B} \times \frac{100}{1} \quad \text{where} \quad \begin{array}{l} B = \text{cost at Beginning of period} \\ E = \text{cost at End of period} \end{array}$$

For purposes of comparison, the period of time over which it is measured has to be standardised, and the period generally used is one year. Though it can be applied anywhere a cost is involved, it is most commonly thought of in connection with the **RPI** *where the year on year change in the RPI is given as the rate of inflation. The table gives the inflation figures (rounded to 1 d.p.) produced from the above table for the RPI.*

So, for 1998 the percentage rate of inflation is

$$\frac{159.5 - 154.4}{154.4} \times \frac{100}{1} = 3.3031$$

The rate of inflation could be negative – it was in the UK in the 1920s – but it is usually positive.

1988	3.3
1989	7.5
1990	7.7
1991	9.0
1992	4.5
1993	1.7
1994	4.1
1995	3.3
1996	2.9
1997	2.8
1998	3.3
1999	2.4
2000	2.0

annuity An annuity is a fixed amount of money paid yearly in return for a sum of money given to the business which is paying the annuity. *People often buy an annuity (from an insurance company) in order to have a regular income after they retire. The size of an annuity is dependent not only upon the sum of money paid, but also upon the age and sex of the person (which is a guide to how long they might live). Thus, a sum of £20 000 might produce an annuity of £2300 for a 55-year-old man (a return of 11.5% per year) but only £2150 for a 55-year-old woman (10.75%).*

arithmetic (the four rules)

the four rules of arithmetic are the operations of **addition, subtraction, multiplication** and **division**.

addition is the operation of combining numbers, each of which represents a separate measure of quantity, so as to produce a number representing the measure of all those quantities together. *If the number is to have any physical meaning, then the quantities must be of the same type.*
Example: Addition shows that 9 people put with 4 people makes 13 people.

total A total is the final number produced by the process of **addition**.
Example: In 9 + 4 = 13 the number 13 is the total.

aggregate ≡ **total**

sum may mean: any process in arithmetic needed to solve a problem
     OR: the process of **addition** on some specified numbers
     OR: the result produced by a process of **addition** (= **total**).
Examples: 'How many sums did you get right?',   'Find the sum of all the numbers from 1 to 9.',   'The sum of 8 and 4 is 12.'

subtraction is the operation of finding a number which gives a measure of the difference in size between two quantities or measures.
Examples: Taking 9 people from a group of 13 leaves 4   (13 − 9 = 4)
$$5°C - (^-3°C) = 8°C$$

difference The difference of two numbers is the result of a **subtraction**.

absolute difference The absolute difference is the **difference** between two numbers ignoring any negative sign in the answer. *It is a measure of the 'distance' between the two numbers when placed on a **number line**.*
Example: The absolute difference of 3 and 5 is 2;   of $^-3$ and 5 it is 8

~ is the symbol meaning that the **absolute difference** is needed. *(3 ~ 5 = 2)*

decomposition is a method of **subtraction** which breaks down (decomposes) the first number in the operation, where necessary, to allow the subtraction to take place. *It works like this:*

$$
\begin{array}{r} 874 \\ -629 \\ \hline \end{array}
\;\text{is}\;
\begin{array}{r} 800 + 70 + 4 \\ 600 + 20 + 9 \\ \hline \end{array}
\;\text{which becomes}\;
\begin{array}{r} 800 + 60 + 14 \\ -\;600 + 20 + \;9 \\ \hline 200 + 40 + \;5 \end{array}
\;\text{written as}\;
\begin{array}{r} 8\;\;^6\rlap{/}7\;\;^14 \\ -6\;\;2\;\;9 \\ \hline 2\;\;4\;\;5 \end{array}
$$

equal addition is a method of **subtraction** which adds the same amount to both numbers, where necessary, to allow the subtraction to take place. *This is the method where the phrase 'borrow and pay back' occurs. It works like this:*

$$
\begin{array}{r} 874 \\ -629 \\ \hline \end{array}
\;\text{is}\;
\begin{array}{r} 800 + 70 + 4 \\ 600 + 20 + 9 \\ \hline \end{array}
\;\begin{array}{c}\text{add 10 to}\\ \text{both numbers}\end{array}\;
\begin{array}{r} 800 + 70 + 14 \\ -\;600 + 30 + \;9 \\ \hline 200 + 40 + \;5 \end{array}
\;\text{written as}\;
\begin{array}{r} 8\;\;7\;\;^14 \\ -6\;\;^3\rlap{/}2\;\;9 \\ \hline 2\;\;4\;\;5 \end{array}
$$

counting on is a method of **subtraction** which finds the **difference** between two numbers by counting on from the smaller to the larger and then adding up all the 'steps' needed. *The steps can be of any convenient size that can be handled mentally. It is sometimes called the 'shopkeeper's method' from the way change used to be given.*
Example: $629 \longrightarrow 630 \longrightarrow 700 \longrightarrow 800 \longrightarrow 874$
                 $+1$ $+70$ $+100$ $+74$ $=245$

complementary addition is a method of **subtraction** that makes a new number from the second number of the operation by: subtracting all its digits from 9; adding this new number to the first number of the operation; adding 1 to the answer; finally subtracting 1 from the 'extra digits'. *Extra digits are all those which go beyond the length of the second number.*

874	874	*adding 1 to the answer makes* 1245
− 629 *change to*	+ 370	*only 'extra digit' is left-hand* 1 *(take off 1 = 0)*
	1244	*and final answer* = 245

35 874 − 629 would first give 36 244 and the extra digits would be 36. Subtracting 1 gives the final answer: 35 245

multiplication is the operation which combines several equal measures of size giving the result as a single number. *With whole numbers multiplication can be seen as equivalent to the addition of several numbers of the same size; the more general case for all numbers is an extension of that. Example: There are 4 rooms with 6 people in each, so in total there are 6+6+6+6 or 6 × 4 people. This extends to cases like 3.28 × 5.74*

$\times$ * are two symbols, both meaning that multiplication is to be done.

product The product is the result given by the operation of **multiplication**. *Example: The product of 1.6 and 7 is 1.6 × 7 which is 11.2*

division is the operation between two numbers which measures how many times bigger one number is than the other. *With whole numbers division can be seen as equivalent to the sharing out of a quantity into a number of equal-sized portions; the general case for all numbers is an extension of that.*

$\div$ / are two symbols, both meaning that **division** is to be done.
Examples: 18 ÷ 6 = 3 and 18/6 = 3

short division is the description used when **division** is done mentally. *Numbers may be written down, but none of the processes are.*

long division is an **algorithm** to deal with **division** for those cases where the numbers are too difficult to work with mentally. *An example is shown on the right.*

$$\begin{array}{r} 67 \\ 23\overline{)1541} \\ 138 \\ \overline{161} \\ 161 \end{array}$$

quotient The quotient is the result given by the operation of **division**. *Example: In 32 ÷ 8 = 4 the quotient is 4*

dividend The dividend is the amount in a **division** operation which is to be shared out, or the number which must be divided into parts. *Example: In 21 ÷ 3 = 7 the dividend is 21*

divisor The divisor is the amount in a **division** operation which must do the dividing. or among which the **dividend** must be shared. *Example: In 18 ÷ 9 = 2 the divisor is 9*

remainder The remainder is the amount left over in a **division** operation when one quantity cannot be divided exactly by another. *Example: In 23 ÷ 4 the answer (quotient) is 5 with a remainder of 3 which is usually written as 5 r 3 or 5 rem 3*

Dividend ÷ Divisor = Quotient *and* Remainder *that cannot be shared out*

calculating aids

tally A tally is a physical record of an amount; usually applied to a system which records the amount as it is being counted. *The best known examples are: notches made on a piece of wood; stones dropped into a receptacle of some sort; marks made on paper, usually grouped in 5's in the way shown on the right* .

$$\text{卌 } || \quad = 7$$
$$|||| \quad = 4$$
$$\text{卌 卌 }|||\backslash = 13$$

abacus An abacus is a device that allows numbers to be shown physically and then manipulated so that calculations may be carried out. *One early form was simply a board covered with dust; another was a table with lines drawn upon it, or grooves, along which counters could be moved. But the form now most associated with the word is a frame holding some wires along which beads can be moved. It is often seen as a child's toy.*

s'choty A s'choty is the Russian form of an **abacus** of the wire and bead type. *There have been different varieties, some of which allowed fractions to be handled. The most general type is shown; the wires are actually curved (upwards in the middle) so that the beads must go to one side or the other. The coloured beads are only a visual aid to counting and the short row repesents either fractions (quarters in this case) or a decimal point. The starting (zero) position is with all beads to the right so the beads to the left in the diagram show the number 317.26*

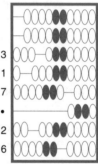

suan pan The suan pan is the Chinese form of an **abacus** of the wire and bead type where the frame is divided internally into two (unequal) sections; the smaller section holds two beads on each wire (each bead counting as a 5) while the larger section has five beads (each counting as a 1). *Only the beads touching the dividing bar count so the number showing is 80 162*

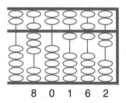

8 0 1 6 2

soroban The soroban is a development of the **suan pan** and mainly used in Japan; it has only one bead on each wire in the smaller section and four in the other. *Again it is only the beads touching the dividing bar which count, so the number showing is 40 925*

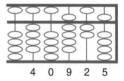

4 0 9 2 5

gelosia multiplication A gelosia is a grid. It was used as the basis of a method of multiplication which was then identified by that name. *The diagram on the right shows 635 × 724. Each separate pair of digits is multiplied together and the answer written in the appropriate square in the way shown with a split between the tens and the units. All these answers are then added up diagonally, remembering to 'carry' a figure where necessary, to produce a final answer of 459 740*

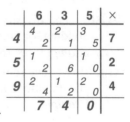

slide rule A slide rule is a device for doing multiplication or division (and other similar operations) by means of two identically graduated strips positioned one beside the other. *The example below shows part of a very simplified slide rule set up to do 1.7 × 3.2, showing 5.44 as the answer. Conversely this same setting of the slide rule is showing 5.44 ÷ 3.2 = 1.7 and together with a knowledge of how to multiply and divide by 10, provides answers to: 17 × 320; 0.17 × 32; 544 ÷ 32; 54.4 ÷ 0.32; and so on.*

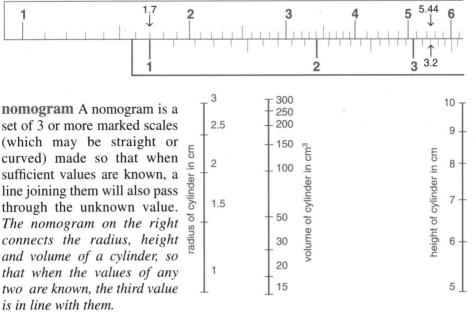

nomogram A nomogram is a set of 3 or more marked scales (which may be straight or curved) made so that when sufficient values are known, a line joining them will also pass through the unknown value. *The nomogram on the right connects the radius, height and volume of a cylinder, so that when the values of any two are known, the third value is in line with them.*

Napier's rods or **bones**, invented to help with multiplication, were based on the **gelosia** method. Each rod was a flat strip with a single digit at the top and nine cells underneath containing the result of multiplying this top digit by 1 to 9 in order from top to bottom. *Rods for 3, 5, 6 and 7 are shown on the right. The index rod merely indicates the multiplier used on each level. Example: Three of the rods are shown put together to form the number 635 (at the top). The shaded areas show the cells used when multiplying 635 by 2, 4 and 7 respectively. Remembering that the figures have to be added diagonally, the three answers are 1270, 2540 and 4445. To do the sum 635 × 724 (as with the gelosia opposite) requires some further ingenuity since the place values are not fixed automatically with these rods.*

calculus is a word used to describe a system of rules of reasoning that is used for doing a certain type of calculation. *There are different types of calculus but the one that is usually meant when the word is used by itself is the* **infinitesimal calculus**.

infinitesimal An infinitesimal quantity is one which is so small that it is almost indistinguishable from zero.

δ (delta) is the symbol used to indicate that a very small value is to be used. *Example: δx, δy etc.*

infinitesimal calculus The infinitesimal calculus is rooted in the idea of considering the effects of **infinitesimal** changes in the values of a function. *An important use of this calculus is in the evaluation and analysis of functions when drawn as graphs. It is generally just called* **calculus**.

differential calculus The differential calculus is that part of **calculus** which is concerned with the rate at which a function is changing. *If the function is shown as a graph, the differential calculus allows the* **gradient** *of the graph at any particular point to be calculated.*

derivative The derivative of a function for some particular value is a measure of EITHER the rate at which the function is changing at that value, OR the **gradient** of the graph of the function at that point. *It is also known as the* **differential coefficient**.
Example: For $f(x) \equiv x^2 + 4x - 7$ *the derivative when x = 3 is 10*

derived function From a given function f(x) the rules of calculus allow another function to be made, known as the derived function and shown as f'(x), which allows a **derivative** of the given function to be calculated for any particular value of x. *Unfortunately, the derived function is sometimes referred to as the* **derivative**.
Example: Given $f(x) \equiv x^2 + 4x - 7 \Rightarrow f'(x) \equiv 2x + 4$ *then, when x = 3 the value of f'(x) is 10 which is the derivative of f(x) when x = 3*

stationary value A stationary value of a function f(x), if it has one, is the value of that function at any point where the **derivative** is zero.

stationary point When the function f(x) is drawn as a graph, stationary points on the curve occur against any x-values which produce a **stationary value**. *That is, the curve will have zero gradient at that point (since the derivative is zero). Such points are a* **maximum**, **minimum** *or* **point of inflection**.

maximum Consider moving along the curve made by the graph of f(x) and passing through a **stationary point**. If the value of f(x) is INCREASING before the stationary point and DECREASING after it, then that stationary point is a maximum. It is a **global maximum** if no greater value of f(x) exists; otherwise it is a **local maximum**.

minimum Consider moving along the curve made by the graph of f(x) and passing through a **stationary point**. If the value of f(x) is DECREASING before the stationary point and INCREASING after it, then that stationary point is a minimum. It is a **global minimum** if no smaller value of f(x) exists; otherwise it is a **local minimum**.

point of inflection If, at a **stationary point**, the curve of f(x) is neither a **maximum** nor a **minimum** then it is a point of inflection.

In the diagram on the right the red dots • are all stationary points. Note there is no global minimum.

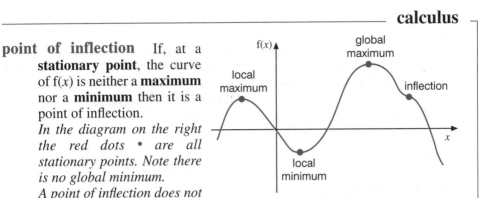

A point of inflection does not have to be at a stationary point. It occurs at any point where the tangent (to the curve) crosses the curve. The commonest example of this is between a maximum and a minimum.

integral calculus The integral calculus is that part of **calculus** which is concerned with summing the values of a function over a particular range. *One of its principal uses is in finding the areas of irregular shapes.*

∫ is the sign used to show that an integral is required to be calculated. *In its most general form it is written as* ∫ f(x) dx, *meaning that the integral is to be done with regard to x. In that form it would be described as an **indefinite integral** since it does not state the range over which it is to be calculated.*

definite integral A definite integral is one in which the range over which the integral is to be calculated is given. *In effect this means that the area under the graph of* f(x) *between the stated end-points* (x = a, x = b) *is being found. The shaded area on the right is given by* ∫$_a^b$ f(x) dx.

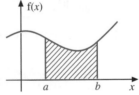

infinity The concept of infinity is of a space, time or quantity that knows no bounds, it goes on for ever and cannot be measured on any practical scale that we know of. *In terms of number, no matter what number may be thought of, it is clear that there must be another number which is bigger, and so on, and so on – to infinity.*

∞ is the symbol used to represent **infinity**.

tends to infinity While infinity as a value cannot ever be reached, we can consider the case of a number (usually *n*) which is growing without restraint and always moving towards infinity. This is expressed as '*n* tends to infinity' and written $n \to \infty$.

limit Consider the sequence of fractions:

$$\frac{1}{1}, \frac{1}{2}, \frac{1}{3}, \frac{1}{4}, - - - - \frac{1}{100}, - - - \frac{1}{1000}, - - - \frac{1}{n}$$

Clearly it can go on for ever with no limit on the value of *n*. Since we CANNOT have $n = \infty$ we must consider what happens as *n* gets indefinitely large, or as *n* **tends to infinity**. Noting that as *n* gets bigger the fraction gets smaller, and as the limit on smallness is zero, we reach the conclusion that, in the limit as *n* tends to infinity the fraction becomes zero. *This is written as shown on the right.*

$$\lim_{n \to \infty} \frac{1}{n} = 0$$

circle A circle is EITHER a **closed curve** (= *a line which curves around and joins up with itself*) in a **plane** (= *a flat surface*), which is everywhere the same distance from a single fixed point, OR it is the shape enclosed by that curve.

Area of circle = $\pi \times$ Radius $\times$ Radius = πr^2

centre The centre of a **circle** is the fixed point from which the distance to the closed curve forming the circle is measured.

radius The radius of a **circle** is the distance from the **centre** to the curve which makes the circle. A radius is any straight line from the centre to the curve.

diameter A diameter of a **circle** is a straight line passing through the **centre** and which touches the curve forming the circle at each of its ends.

semicircle A semicircle is one half of a **circle** made by cutting along a diameter.

circumference The circumference of a **circle** is the distance measured around the curve which makes the circle.

Circumference = $\pi \times$ Diameter = πd

chord A chord of a **circle** is any straight line drawn across a circle, beginning and ending on the curve making the circle. *A chord which passes through the centre is also a diameter. A line extending beyond the circle is a secant.*

arc An arc of a **circle** is any piece of the curve which makes the circle.

$$\text{Length of arc} = \frac{\text{Angle (\textit{in degrees}) of arc at centre}}{360} \times \text{Circumference of full circle}$$

sector A sector of a **circle** is the shape enclosed between an **arc** and the two radii at either end of that arc.

$$\text{Area of sector} = \frac{\text{Angle (\textit{in degrees}) of sector at centre}}{360} \times \text{Area of full circle}$$

segment A segment of a **circle** is the shape enclosed between a **chord** and one of the **arcs** joining the ends of that chord.

$$\text{Area of segment} = \left(\frac{\pi\theta}{360} - \frac{1}{2}\sin\theta \right) \times r^2 \quad \text{where}$$

θ (*in degrees*) is the angle of the segment at the centre
r is the radius of the circle

major } { **arc**
sector
minor } { **segment**

When one **arc, sector** or **segment** is made in a circle then the remainder of the circle makes another arc, sector or segment. The LARGER is known as the MAJOR, the SMALLER as the MINOR arc, sector or segment.

concentric circles are two or more circles which have been drawn using the SAME position for their centres.

eccentric circles are two or more circles which have been drawn using DIFFERENT positions for their centres. *Usually, for two circles, one is completely inside the other, or else there is some area which is common to all the circles.*

annulus An annulus is a shape like a ring which is formed by the space enclosed between two **concentric circles**.

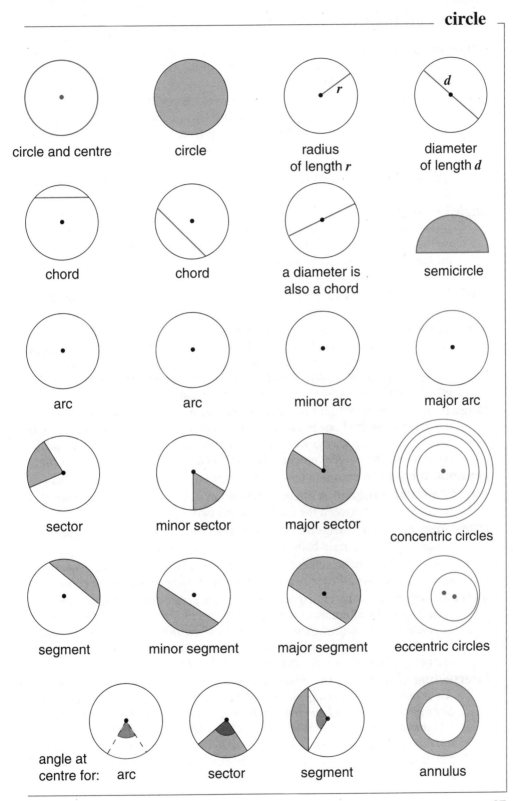

circle and centre

circle

radius
of length *r*

diameter
of length *d*

chord

chord

a diameter is
also a chord

semicircle

arc

arc

minor arc

major arc

sector

minor sector

major sector

concentric circles

segment

minor segment

major segment

eccentric circles

angle at
centre for: arc

sector

segment

annulus

circles and their properties

angle properties of circles is a collection of theorems which give the relationships between various parts of a circle (chord, segment, etc.) and angles associated with them.

subtended angle Given three distinct points A, B and C (which are not in a straight line), the subtended angle of any two of the points at the third, is the angle formed between the lines drawn from the first two points to the third. *The subtended angle of points A and C at B would be the angle formed between the lines BA and BC.*
Example: The angle subtended by the diameter of the Moon at any point on the Earth is about half a degree.

angle in a segment The angle in a segment is the angle formed between the two lines drawn from the ends of the **chord** making the **segment** to any point on the circumference of that segment. *It is the angle subtended at the point on the circumference by the chord. In any given segment all the subtended angles are the same size.*

angle in a semicircle In any semicircle the angle **subtended** by the diameter at any point on the circumference is a right angle.

angle at the centre The angle at the centre of a circle is the one formed between the two radii drawn from two points on the circumference. *It is the angle subtended at the centre by the chord defined by those two points. Given any chord, the angle it subtends at the centre is twice the angle in the segment which is on the same side of the chord as the centre.*

tangent A tangent to a circle is a line which, no matter how far it is extended, touches the circle at one point only. *From any one fixed point outside a circle two tangents can always be drawn to that circle. The radius drawn at the point where the tangent touches the circle is at right angles to the tangent.*

common tangent A common tangent is a **tangent** that touches two circles.

direct common tangent A direct common tangent is a **common tangent** that DOES NOT pass between the centres of the two circles.

transverse common tangent A transverse common tangent is a **common tangent** that DOES pass between the centres of the two circles.

secant A secant is a line which cuts across a circle at two points. *A tangent can be considered a special case of a secant in which the two points are coincident.*

alternate segments Any **chord** drawn in a circle creates two segments, and one is said to be the alternate of the other. *When a chord is drawn from the point of contact of a tangent then the angle between the tangent and the chord, measured on ONE side of the chord, is equal to the angle in the alternate segment, which lies on the OTHER side of the chord.*

intersecting chords are two **chords** drawn in the same circle which cross at some point. *The crossing point may be outside the circle and require the lines of the chords to be extended. If the two chords are labelled as AB and CD and they cross at O, then $OA \times OB = OC \times OD$. If the point O falls outside the circle, then the lines OAB and OCD are secants. In the particular case where O is outside the circle with OT a tangent and OAB a secant to that circle then $OA \times OB = OT^2$*

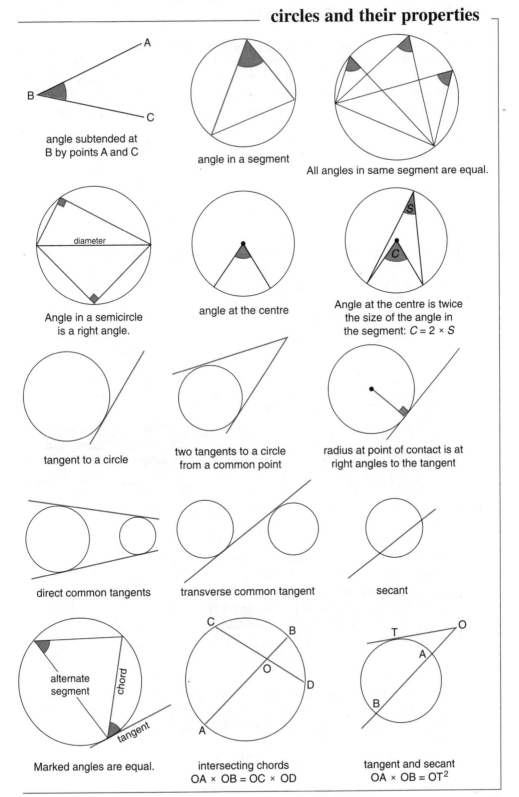

angle subtended at
B by points A and C

angle in a segment

All angles in same segment are equal.

diameter

Angle in a semicircle
is a right angle.

angle at the centre

Angle at the centre is twice
the size of the angle in
the segment: $C = 2 \times S$

tangent to a circle

two tangents to a circle
from a common point

radius at point of contact is at
right angles to the tangent

direct common tangents

transverse common tangent

secant

alternate
segment

chord

tangent

Marked angles are equal.

intersecting chords
$OA \times OB = OC \times OD$

tangent and secant
$OA \times OB = OT^2$

circles related to other shapes

circumcircle A circumcircle to any **polygon** is the circle drawn around the OUTSIDE of that polygon which touches ALL of its vertices. *Since it is necessary for the circle to touch every vertex of the polygon, it is not possible to draw a cirmcumcircle for every polygon, but it is always possible for regular polygons and for any triangle. The position of its centre is called the* **circumcentre**. *Examples:*

circumcircle to a
regular hexagon

circumcircle to an
irregular hexagon

circumcircles to triangles

incircle An incircle to any **polygon** is the circle drawn on the INSIDE of that polygon which touches ALL its edges. *Each edge is a tangent to the incircle. As with the circumcircle, not every polygon has an associated incircle, but every regular polygon has one, and so do all triangles. The position of its centre is called the* **incentre**. *Examples:*

incircle to a
regular pentagon

incircle to an
irregular pentagon

incircles to triangles

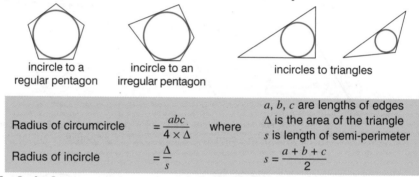

Radius of circumcircle	$= \dfrac{abc}{4 \times \Delta}$	where	a, b, c are lengths of edges Δ is the area of the triangle s is length of semi-perimeter
Radius of incircle	$= \dfrac{\Delta}{s}$		$s = \dfrac{a + b + c}{2}$

escribed circles An escribed circle to any **polygon** is the circle which touches the outside of one edge of the polygon and also the two adjacent edges extended as necessary. *Escribed circles can be drawn on any polygon, but they are most commonly associated with triangles.*

Using the same notation as above:
The three escribed circles for a triangle with edges a, b, c have radii of r_a r_b r_c respectively

$$r_a = \frac{\Delta}{s - a} \qquad r_b = \frac{\Delta}{s - b} \qquad r_c = \frac{\Delta}{s - c}$$

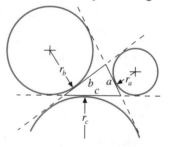

Apollonius' circle Given two fixed points A and B then the **locus** of a point P moving in the plane of A and B, in such a way that the value of AP ÷ BP is constant, is a circle. *In the drawing on the right the distance AB is 15 mm and AP ÷ BP = 2*

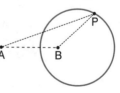

cyclic quadrilateral A cyclic quadrilateral is a **quadrilateral** around which a circle can be drawn to pass through all of its vertices. *It would be the* **circumcircle** *for that quadrilateral. The opposite vertex angles of a cyclic quadrilateral add up to 180 degrees. Rectangles and isosceles trapeziums are* ALWAYS *cyclic quadrilaterals; the kite and irregular quadrilaterals sometimes are; the rhombus, parallelogram and arrowhead can* NEVER *be. Examples:*

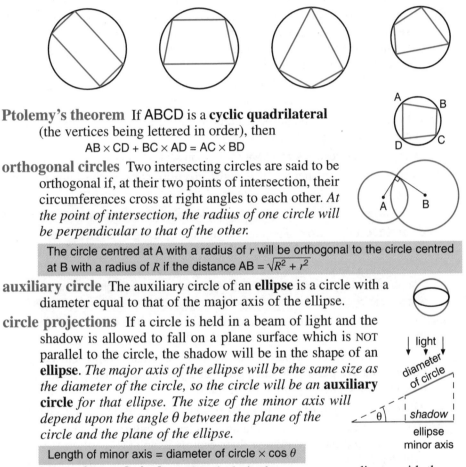

Ptolemy's theorem If ABCD is a **cyclic quadrilateral** (the vertices being lettered in order), then
$$AB \times CD + BC \times AD = AC \times BD$$

orthogonal circles Two intersecting circles are said to be orthogonal if, at their two points of intersection, their circumferences cross at right angles to each other. *At the point of intersection, the radius of one circle will be perpendicular to that of the other.*

> The circle centred at A with a radius of r will be orthogonal to the circle centred at B with a radius of R if the distance AB = $\sqrt{R^2 + r^2}$

auxiliary circle The auxiliary circle of an **ellipse** is a circle with a diameter equal to that of the major axis of the ellipse.

circle projections If a circle is held in a beam of light and the shadow is allowed to fall on a plane surface which is NOT parallel to the circle, the shadow will be in the shape of an **ellipse**. *The major axis of the ellipse will be the same size as the diameter of the circle, so the circle will be an* **auxiliary circle** *for that ellipse. The size of the minor axis will depend upon the angle θ between the plane of the circle and the plane of the ellipse.*

> Length of minor axis = diameter of circle × cos θ

transformations of circles If a circle is drawn on a coordinate grid, then a **transformation** can be applied to all the points which define that circle to produce a new shape. *This is usually done using* **matrices**. *A simple application of this is to to transform a circle (by means of a one-way stretch) into an ellipse. Example:*

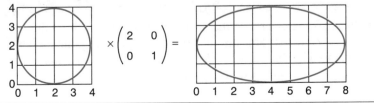

cones, cylinders and spheres

cone A cone is the three-dimensional shape formed by a straight line when one end is moved around a **simple closed curve**, while the other end of the line is kept fixed at a point which is not in the plane of the curve.

vertex The vertex of a **cone** is the fixed point used in making it.

base The base of a **cone** is the simple closed curve used in making it.

circular cone A circular cone is a **cone** made using a circle as its **base**.

right circular cone A right circular cone is a **cone** made using a circle as its **base** and with its **vertex** placed on a line passing through the centre of the base and perpendicular to the plane of the base. *It is what is usually meant when only the word 'cone' is used.*

oblique circular cone An oblique circular cone is a NON-**right circular cone**. *The vertex is not placed over the centre of the base.*

perpendicular height The perpendicular height of a **cone** is the distance of its **vertex** above the plane of its **base**.

> Volume of any cone = Area of base × Perpendicular height ÷ 3

slant height The slant height of a **right circular cone** is the length of any straight line from the circumference of its **base** to the **vertex**.

> Slant height $= \sqrt{r^2 + h^2}$ where r is radius of base h is perpendicular height

curved surface of a cone The curved surface of a **right circular cone** is the sector which could be bent around (until the edges meet) to form the cone.

> Radius of circle to make sector = Slant height of cone
> Angle of sector (*in degrees*) = 360 × Base radius ÷ Slant height of cone
> Area of sector = π × Base radius × Slant height of cone

frustum of a cone The frustum of a **cone** is the part of the cone cut off between the **base** and a plane which is parallel to the base.

cylinder A cylinder is formed by using two identical **simple closed plane curves** that are parallel to each other and joining up corresponding points on each of the curves with straight lines.

ends The ends of a **cylinder** are the two simple closed curves used to make it.

right circular cylinder A right circular cylinder is a **cylinder** in which the **ends** are circles and the line joining their centres is an axis of symmetry of the cylinder. *It is what is usually meant when only the word 'cylinder' is used.*

> Volume of a right circular cylinder = Area of one end × Distance between ends

curved surface of a cylinder The curved surface of a **right circular cylinder** is the rectangle which could be bent around (until two opposite edges meet) to fit the two circular ends and so form the complete cylinder.

> Area of curved surface $= 2\pi rh$ where r is radius of end h is height (or length)

sphere A sphere is EITHER the shape of a surface in three dimensions which is everywhere the same distance from a single fixed point OR it is the solid shape enclosed by that surface. *The balls used to play most games are spheres.*

hemisphere A hemisphere is one half of a **sphere**.

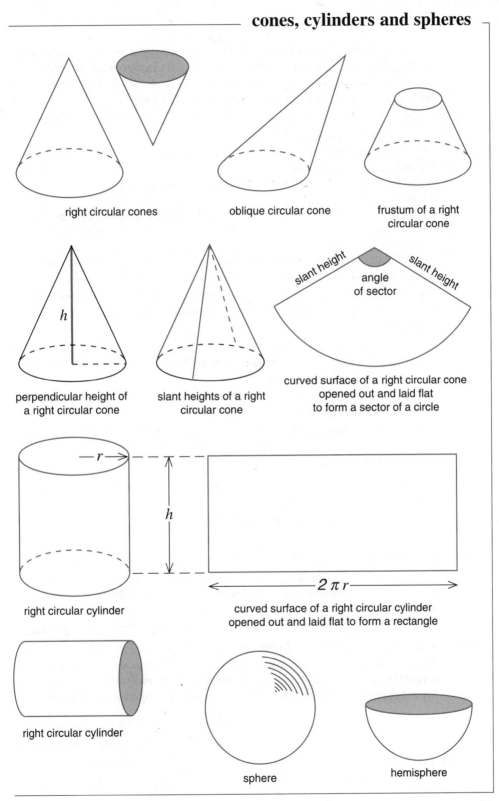

right circular cones

oblique circular cone

frustum of a right
circular cone

h

perpendicular height of
a right circular cone

slant heights of a right
circular cone

slant height

angle
of sector

slant height

curved surface of a right circular cone
opened out and laid flat
to form a sector of a circle

r

h

right circular cylinder

$2\pi r$

curved surface of a right circular cylinder
opened out and laid flat to form a rectangle

right circular cylinder

sphere

hemisphere

conic sections is the general name given to the four types of curve that can be produced by the section of a **cone** as it is sliced through by a straight cut at various angles. *As the angle at which the cut is made changes, the curves produced are the* **circle, ellipse, parabola** *and* **hyperbola**.

focus A focus is a fixed point used in the drawing of any of the **conic sections**.

ellipse An ellipse is the **locus** of a point which moves in such a way that its distances from two **foci** add together to a constant amount. *Once the two foci are fixed and the constant amount is decided then only* ONE *ellipse is possible. It may be drawn as a graph using the equation*

$$\frac{x^2}{a^2} + \frac{y^2}{b^2} = 1 \quad \text{where } a, b \text{ are constant values affecting the size}$$

Its area is given by πab

Its perimeter is difficult to calculate exactly, but Ramanujan's formula gives a very accurate approximation: $\pi[3(a + b) - \sqrt{(a + 3b)(3a + b)}]$.

major axis The major axis of an **ellipse** is the straight line drawn through the two foci with each end of the line touching the ellipse. *It is a line of symmetry.*

minor axis The minor axis of an **ellipse** is the longest straight line that can be drawn at right angles to the **major axis** with each end of the line touching the ellipse. *It is a line of symmetry.*

a, b are the symbols used to give the size of an ellipse, being half the lengths of the **major** and **minor** axes respectively.

eccentricity The eccentricity of an **ellipse** is a measure of how much it varies from a circle. *Its value is given by the formula* $\sqrt{1 - \frac{b^2}{a^2}} \quad (a > b)$

circle A circle can be considered as a special case of the **ellipse**, with the **major** and **minor** axes equal in length ($a = b$). *It is an ellipse with only one focus and its eccentricity is zero. It may be drawn as a graph using the equation*

$$x^2 + y^2 = a^2 \quad (a \text{ is the radius})$$

directrix A directrix is a fixed straight line used in drawing the conic curves.

parabola A parabola is the **locus** of a point that moves in such a way as to be always the same distance from a **focus** as it is from a **directrix**. *Once the focus and the directrix are fixed, only one parabola is possible. It may be drawn as a graph using the equation* $\quad y^2 = 4ax \quad (a \text{ is a constant})$

hyperbola A hyperbola is the **locus** of a point that moves in such a way as to make its distance from a **focus** always greater than its distance from a **directrix** in some ratio. *Once the focus, directrix and ratio are fixed,* TWO *hyperbolas are possible. They may be drawn as graphs using the equation*

$$\frac{x^2}{a^2} - \frac{y^2}{b^2} = 1 \quad (a, b \text{ are constant values affecting the size})$$

rectangular hyperbola A rectangular hyperbola is the **hyperbola** produced by a graph having the equation $\quad xy = k \quad (k \text{ is a constant})$

Slicing through a cone in these directions:

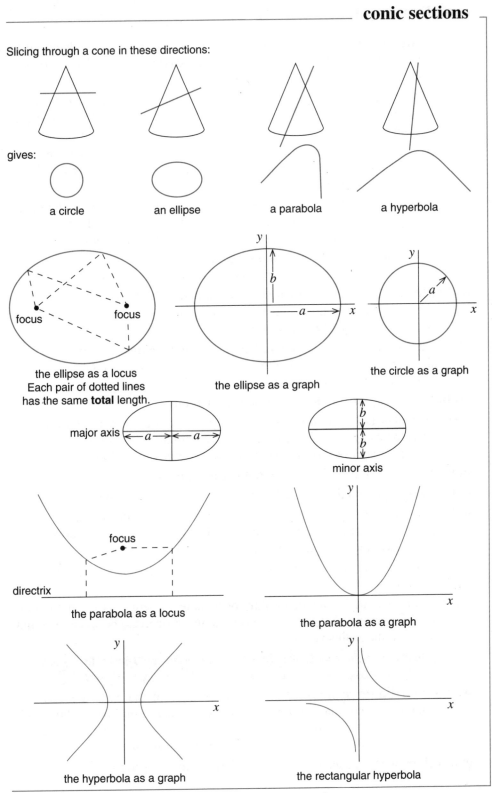

gives:

a circle   an ellipse   a parabola   a hyperbola

focus   focus

the ellipse as a locus
Each pair of dotted lines
has the same **total** length.

the ellipse as a graph

the circle as a graph

major axis

minor axis

focus

directrix

the parabola as a locus

the parabola as a graph

the hyperbola as a graph

the rectangular hyperbola

coordinate systems are used to give the position of a point by placing it in relation to some other fixed positions. *The fixed positions might be points, lines or planes, depending on the system.*

Cartesian coordinates give the position of a point in two-dimensional space by stating its shortest distances from 2 fixed reference lines set at right angles to each other. *The distances may be given as positive or negative values.*

axes The axes are the 2 fixed lines in the **Cartesian coordinate** system. *They are usually identified separately as the x-axis and the y-axis and are placed at right angles to each other.*

ordered pair An ordered pair is the two numbers, written in a particular order, needed to give the position of a point in the **Cartesian coordinate** system. *The convention is to give the x-number first, and the y-number second.*

origin The origin in the **Cartesian coordinate** system is the point where the two **axes** cross. *It is the point identified by the ordered pair (0,0).*

ordinate The ordinate of a point in **Cartesian coordinates** is its distance from the x-axis, as measured on the y-axis. *It is the value of the second number in the ordered pair for that point.*

abscissa The abscissa of a point in **Cartesian coordinates** is its distance from the y-axis, as measured on the x-axis. *It is the value of the first number in the ordered pair for that point.*

rectangular coordinates ≡ **Cartesian coordinates**

three-dimensional coordinates give the position of a point in 3-dimensional space by using 3 fixed reference lines. *The 3rd axis, identified as the z-axis, is at right angles to both the x- and y- axes. For this, an ordered triple is used.*

pole A pole is a fixed point used in **polar coordinates**.

polar axis A polar axis is a fixed line, one end of which is a **pole**.

polar coordinates give the position of a point in two-dimensional space by stating its DISTANCE from a **pole** and the size of the ANGLE between the **polar axis** and a line drawn from the point to the pole.

radius vector The radius vector is the line joining the **pole** and the point whose position is being given.

(r, θ) is the symbol for showing **polar coordinates**, where r is the length of the **radius vector** and θ is the angle (in radians or degrees) between the **polar axis** and the radius vector.

> To change polar coordinates (r, θ) into Cartesian coordinates (x, y) use:
> $$x = r \cos \theta \qquad\qquad y = r \sin \theta$$

world coordinate system Positions on the face of the earth are given by reference to an imaginary coordinate system based on lines of **longitude** and lines of **latitude**. *Longitude corresponds to the x-numbers, and latitude to the y-numbers in the Cartesian coordinate system.*

grid references as used on maps are a **Cartesian coordinate** system.

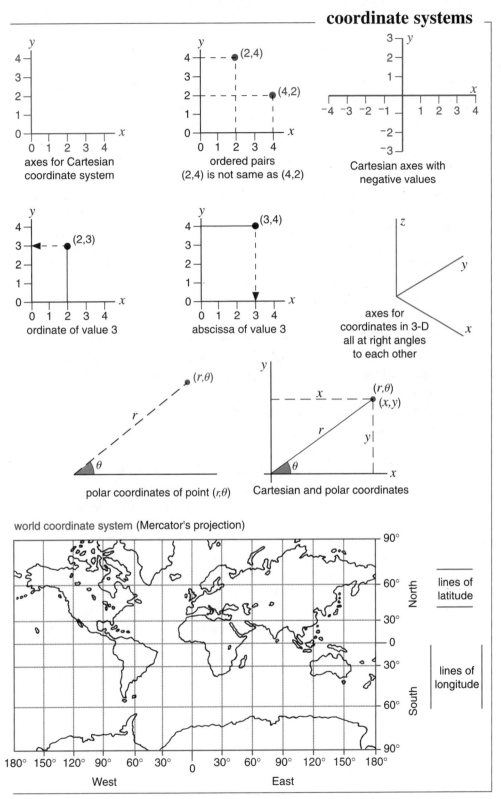

axes for Cartesian
coordinate system

ordered pairs
(2,4) is not same as (4,2)

Cartesian axes with
negative values

ordinate of value 3

abscissa of value 3

axes for
coordinates in 3-D
all at right angles
to each other

polar coordinates of point (r,θ)

Cartesian and polar coordinates

world coordinate system (Mercator's projection)

90°
60° North — lines of latitude
30°
0
30° South — lines of longitude
60°
90°

180° 150° 120° 90° 60° 30° 0 30° 60° 90° 120° 150° 180°
West   East

37

plane curve A plane curve is a **curve** whose entire length lies within a single flat surface or **plane**.

closed curve A closed curve is a **curve** which joins up with itself and has NO end points. *No beginning or ending can be identified.*

simple closed curve A simple closed curve is a **closed curve** which does not cross itself at any point.

arc An arc is part of a **curve**. *It must have two end points, though these might merely be marked to show the arc as part of a bigger curve.*

tangent A tangent to a **curve** is a straight line that touches the curve at a point. *The tangent will have the same gradient as the curve at that point.*

locus A locus is the line of a path along which a point moves so as to satisfy some given conditions. *A locus is usually a curve but does not have to be. Example: A point that moves so that it is always the same distance from another (fixed) point will follow a locus that will be the shape of a circle.*

spiral A spiral is the **locus** of a point moving in the plane of, and around, another (fixed) point while continuously increasing its distance from that fixed point.

Archimedes' spiral is the **spiral** which can be drawn, using **polar coordinates**, from the equation $r = a\theta$ $\qquad$ a is a constant $\neq 0$, θ is the angle

hyperbolic spiral A hyperbolic spiral is the **spiral** which can be drawn, using **polar coordinates**, from the equation
$r = a \div \theta$ $\qquad$ a is a constant $\neq 0$, θ is the angle

equiangular spiral An equiangular spiral is the **spiral** which can be drawn, using **polar coordinates**, from the equation
$r = ae^{k\theta}$ $\qquad$ a, k are constants $\neq 0$, θ is the angle
$\qquad\qquad\qquad$ ($e \approx 2.7182818$)

helix A helix is the shape drawn on the curved surface of a cylinder or cone by a point which moves along the surface at a constant angle. *Two common examples of a helix are a circular staircase and a corkscrew.*

curve of pursuit A curve of pursuit is the line followed by one object moving directly towards another object which is also moving. *The simplest case, when both are moving along the same straight line, is not usually considered.*

catenary A catenary is the curve formed by a heavy uniform string or cable which is hanging freely from two end points. *It is most commonly seen in the overhead cables used to transmit electricity.*

ruled curve A ruled curve is a recognisable curve which is produced by drawing only straight lines. *The smoothness of the curve depends upon the arrangement and number of lines drawn.*

envelope An envelope is the curve which appears as the outline resulting from drawing whole families of other curves.

asymptote An asymptote to a curve is a straight line to which the curve continuously draws nearer but without ever touching it.

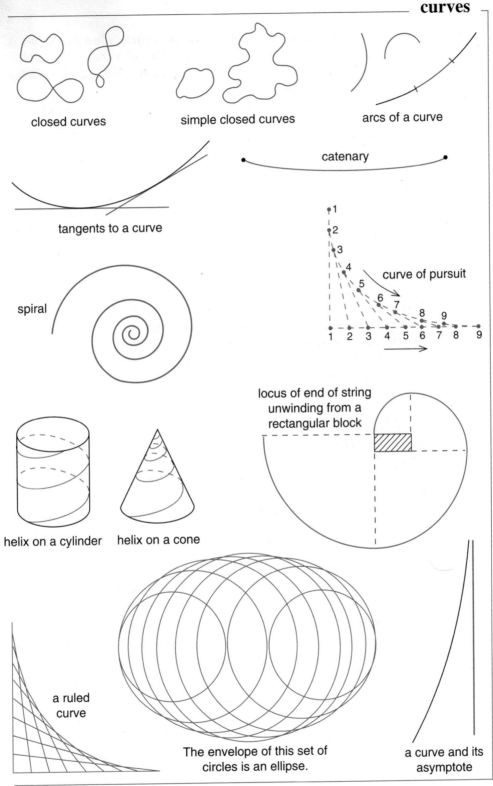

closed curves

simple closed curves

arcs of a curve

catenary

tangents to a curve

curve of pursuit

spiral

locus of end of string
unwinding from a
rectangular block

helix on a cylinder   helix on a cone

a ruled
curve

The envelope of this set of
circles is an ellipse.

a curve and its
asymptote

cycloid A cycloid is the **locus** of a single point on a circle when that circle is rolled along a straight line. *A complete turn of the circle makes one arch.*

Area between one arch and the line = $3\pi a^2$ Length of one arch = $8a$ (a = radius)

cusp A cusp is a point on a curve where the curve makes a complete reversal of direction. *The two parts of the curve share the same tangent at that point.*

parametric equations are those that link two (or more) variables, not directly, but by expressing them in terms of equations using another variable that is common to all the equations. *Examples are given below for the epicycloid and the hypocycloid where coordinates are given in terms of the variable* θ.

epicycloid An epicycloid is the **locus** of a single point on a circle when that circle is rolled around the OUTSIDE of another circle known as the base circle. *The exact shape of this curve, and the number of cusps it has, will be determined by the relative sizes of the rolling circle and the base circle.*

The x,y coordinates needed to plot an epicycloid are given by the **parametric equations**:

$x = a(n \cos \theta - \cos n\theta)$ $y = a(n \sin \theta - \sin n\theta)$

θ may take any values (normally in the range $0°$ to $360°$)
If n is a whole number, then there will be $(n - 1)$ cusps.
The value of a changes the size of the curve.

hypocycloid A hypocycloid is the **locus** of a single point on a circle when that circle is rolled around the INSIDE of another circle known as the base circle. *The exact shape of this curve, and the number of cusps it has, will be determined by the relative sizes of the rolling circle and the base circle.*

The x,y coordinates needed to plot a hypocycloid are given by the **parametric equations**:

$x = a(n \cos \theta + \cos n\theta)$ $y = a(n \sin \theta - \sin n\theta)$

The same remarks apply as for the epicyloid equations above except that in this case there will be $(n + 1)$ cusps.

cardioid A cardioid is an **epicycloid** having only ONE **cusp**. *In the equations for the epicycloid, n = 2. It is the locus drawn when the rolling circle and the base circle are the same size.*

Area enclosed by cardioid = $6\pi a^2$ Perimeter length = $16a$

nephroid A nephroid is an **epicycloid** having only TWO **cusps**. *In the equations for the epicycloid, n = 3. The rolling circle is half the size of the base circle.*

Area enclosed by nephroid = $12\pi a^2$ Perimeter length = $24a$

deltoid A deltoid is a **hypocycloid** having only THREE **cusps**. *In the equations for the hypocycloid, n = 2. The rolling circle is $\frac{1}{3}$rd the size of the base circle.*

Area enclosed by deltoid = $2\pi a^2$ Perimeter length = $16a$

astroid An astroid is a **hypocycloid** having only FOUR **cusps**. *In the equations for the hypocycloid, n = 3. The equations can be rewritten more simply as:*

$x = 4a \cos^3 \theta$ $y = 4a \sin^3 \theta$
Area enclosed by astroid = $\frac{3}{8}\pi a^2$ Perimeter length = $6a$

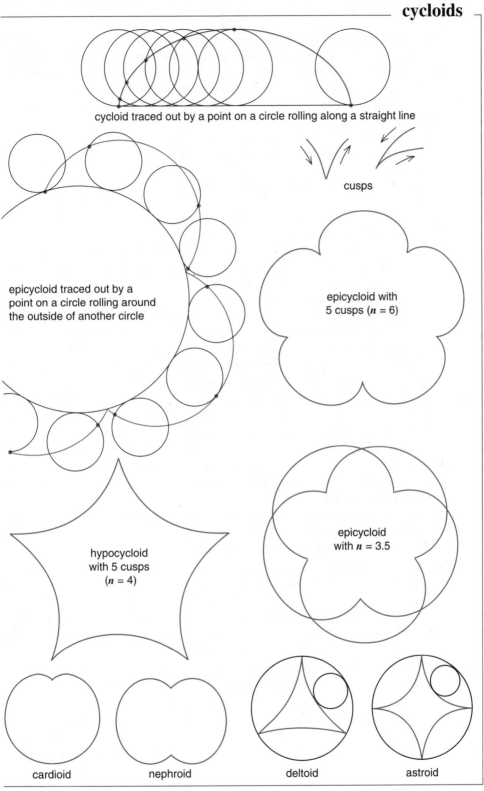

cycloid traced out by a point on a circle rolling along a straight line

cusps

epicycloid traced out by a point on a circle rolling around the outside of another circle

epicycloid with 5 cusps ($n = 6$)

hypocycloid with 5 cusps ($n = 4$)

epicycloid with $n = 3.5$

cardioid

nephroid

deltoid

astroid

eponym An eponym is EITHER the name of a person, factual or fictitious, which is used to form a word or phrase identifying a particular thing OR it is the thing itself. *The person's name is usually that of the one first associated in some way with whatever is being identified. The name may, or may not, be that of the originator. Other eponyms are given under separate topics.*

Eratosthenes' sieve is an **algorithm** for finding **prime numbers**. First write down as many numbers as required to be searched, in order, starting with 1 and not missing any out. Cross out 1. Leave 2 and cross out every 2nd number (4, 6, 8 etc). Leave 3 and cross out every 3rd number (6, 9, 12 etc). 4 is already crossed out, so leave 5 and cross out every 5th number (10, 15 etc). When complete the numbers left are the prime numbers.

~~1~~	2	3	~~4~~	5	~~6~~
7	~~8~~	~~9~~	~~10~~	11	~~12~~
13	~~14~~	~~15~~	~~16~~	17	~~18~~
19	~~20~~	~~21~~	~~22~~	23	~~24~~
~~25~~	~~26~~	~~27~~	~~28~~	29	~~30~~
31	~~32~~	~~33~~	~~34~~	~~35~~	~~36~~
37	~~38~~	~~39~~	~~40~~	41	~~42~~
43	~~44~~	~~45~~	~~46~~	47	~~48~~
~~49~~	~~50~~	~~51~~			

Euclid's algorithm is a method to find if two numbers have a common factor (other than 1), and its value if there is.

> Consider two numbers M and N with $M > N$
> Let $P = M$ and $Q = N$
> → Divide P by Q and find the remainder R
> If $R = 0$ then Q is a factor of M and N
> If $R = 1$ then M and N have only 1 as a common factor
> If $R > 1$ then put $P = Q$ and $Q = R$ and go to →

Fermat's last theorem is that the equation $x^n + y^n = z^n$ has no solutions in whole numbers for x, y and z if $n > 2$ and $x,y,z > 1$, and it is one of the most famous theorems in mathematics. It was really a conjecture since, although Fermat claimed to have a proof, it was not properly proved until 1994.

Fermat's problem is to find the point F in any triangle ABC, that makes the total of the distances AF + BF + CF to be the least. It is also known as **Steiner's problem**.

Gaussian integer A Gaussian integer is a **complex number** in which both the real and imaginary parts are whole numbers.
Examples: 3 + 8i and 10 + 2i are Gaussian integers; 7.6 + 5i is not.

Goldbach's conjecture is that every even number from 4 onwards can be made by adding two **prime numbers**. There are often several possibilities.
Examples: 6 = 3 + 3; 10 = 3 + 7 or 5 + 5; 22 = 3 + 19 or 5 + 17 or 11 + 11

Hamiltonian walk A Hamiltonian walk is a path traced out on a **topological graph** which visits every vertex once and once only – except possibly for the start and finish which might be on the same vertex.

Heronian triangle A Heronian triangle is a triangle whose three edge lengths and its area are all **rational numbers**.

Jordan curve is another name for a **simple closed curve**.

Mersenne primes are those **prime numbers** which can be made from the expression $2^n - 1$. It only works when n itself is prime but even then there are times when it does not work. For instance, it works when $n = 2, 3, 5$ or 7 but not when $n = 11$ or 23, as well as many other prime values.

Mobius band or **strip** A Mobius band is made by taking a rectangular strip of paper like that shown as ABCD and fastening the two shorter edges together (AB and CD) but BEFORE FASTENING giving the strip a half twist so that A fastens to D and B to C. The strange property of this band is that it now has only 1 edge and 1 side.

Pascal's triangle is an array of numbers in the shape of a triangle, having a 1 at the top and also at the ends of each line. All the other numbers are made by adding the pair of numbers closest to them in the line above.

Examples: *1 + 4*   *4 + 6*   *6 + 4*   *4 + 1*
              = 5   *= 10*   *= 10*   *= 5*

```
              1
            1   1
          1   2   1
        1   3   3   1
      1   4   6   4   1
    1   5  10  10   5   1
  1   6  15  20  15   6   1
```

Perigal's dissection is a visual way of illustrating **Pythagoras' theorem**. In the diagram on the right, cutting out and moving the necessary pieces shows how the square drawn on the hypotenuse AB is made up of the (red) square drawn on the shorter edge AC plus the square drawn on the other edge BC which is divided into the four numbered quadrilaterals.

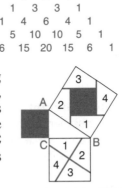

Pick's theorem A grid of dots is marked out in a square array, then a polygon is drawn on this grid by joining up dots with straight lines to make the edges of the polygon, with none of these edges crossing.

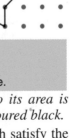

Let the number of dots inside the polygon = *I*
Let the number of dots on the boundary of the polygon = *B*
Then the area of the polygon is ($\frac{1}{2}$ *B* + *I* – 1) × area of a unit square.

Example: In the polygon drawn above, **B** = 10   **I** = 4, *so its area is* (5 + 4 – 1 =) 8 *times the area of the unit square which is coloured black.*

Pythagorean triplets are groups of three numbers (*a*, *b*, *c*) which satisfy the equation $a^2 + b^2 = c^2$ *Basic triplets can be easily made this way:*

Choose two numbers *m*, *n* (*m* > *n*) which have NO common factors (except 1)
Then $a = m^2 - n^2$ $b = 2mn$ $c = m^2 + n^2$

From any basic triplet others can be made by multiplying each of its three numbers by some constant. So (3, 4, 5) gives (6, 8, 10) or (21, 28, 35) etc.

Zeno's paradoxes are concerned with motion and some apparent impossibilities. They were important in the development of mathematical thinking. There are four paradoxes, but the best known is that of Achilles and the tortoise. Achilles races against a tortoise and, since Achilles is clearly the faster of the two, gives the tortoise a good start. Now as soon as Achilles gets to the point at which the tortoise started, the tortoise has moved, and this is repeated over and over. How can Achilles ever catch up with the tortoise?

factors, multiples and primes

Number, as used in this section, means only a **positive whole number.**

factor A factor is a **number** which divides exactly into another **number**. *1 is a factor of every number and every number is a factor of itself. A number can have several factors as shown in the table opposite.*

Examples: 1 is a factor of 5;   3 is a factor of 6;   4 is a factor of 12
7 is a factor of 7;   2 and 17 are factors of 68

proper factors The proper factors of a number are all of its **factors** EXCEPT for the number itself.

Examples: The factors of 12 are 1, 2, 3, 4, 6 and 12, but its proper factors are only 1, 2, 3, 4 and 6
The proper factors of 20 are 1, 2, 4, 5 and 10

proper divisors ≡ proper factors

common factors are those **factors** shared by two (or more) numbers.

Example: 12 has 1, 2, 3, 4, 6, 12 as factors
18 has 1, 2, 3, 6, 9, 18 as factors
So the common factors of 12 and 18 are 1, 2, 3 and 6

highest common factor (hcf) The highest common factor of two (or more) numbers is the **common factor** of all those numbers which has the greatest value. *In some cases the hcf may be 1 or one of the actual numbers.*

Examples: From the previous example, the hcf of 12 and 18 is 6
The hcf of 12 and 17 is 1   The hcf of 5,15, 30 is 5

prime number A prime number is a number having two, and only two, **factors.** *The list opposite gives the first 160 prime numbers. The list could continue for ever. Note that 1 is* NOT *a prime number since it only has one factor.*

prime factors The prime factors of a number are all those **factors** of the number which are themselves **prime numbers**. *A prime number has only one prime factor: itself.*

Examples: All the factors of 12 are 1, 2, 3, 4, 6 and 12 but its only prime factors are 2 and 3

composite number A composite number must have three or more **factors.** *It cannot be 1 or a prime number.*

Examples: 4, 6, 8, 9, 10, 12, 15 are all composite numbers.

multiple A multiple is a number made by multiplying together two other numbers. *A number is a multiple of any of its factors.*

Example: 12 is a multiple of 2 since 12 = 2 × 6

lowest common multiple (lcm) The lowest (or least) common multiple of two (or more) numbers is the smallest possible number into which ALL of them will divide exactly.   *Example: The lcm of 3, 4 and 8 is 24*

fundamental theorem of arithmetic Any composite number can be made by multiplying together a set of prime numbers, and this can be done in only one way. *The reordering of the prime numbers is not considered as different.*

Examples: 12 = 2 × 2 × 3 which, in index form is $2^2 × 3$
126 = 2 × 3 × 3 × 7 or 2 × 3^2 × 7

Number and factors

Number	Factors
1	1
2	1, 2
3	1, 3
4	1, 2, 4
5	1, 5
6	1, 2, 3, 6
7	1, 7
8	1, 2, 4, 8
9	1, 3, 9
10	1, 2, 5, 10
11	1, 11
12	1, 2, 3, 4, 6, 12
13	1, 13
14	1, 2, 7, 14
15	1, 3, 5, 15
16	1, 2, 4, 8, 16
17	1, 17
18	1, 2, 3, 6, 9, 18
19	1, 19
20	1, 2, 4, 5, 10, 20
21	1, 3, 7, 21
22	1, 2, 11, 22
23	1, 23
24	1, 2, 3, 4, 6, 8, 12, 24
25	1, 5, 25
26	1, 2, 13, 26
27	1, 3, 9, 27
28	1, 2, 4, 7, 14, 28
29	1, 29
30	1, 2, 3, 5, 6, 10, 15, 30
31	1, 31
32	1, 2, 4, 8, 16, 32
33	1, 3, 11, 33
34	1, 2, 17, 34
35	1, 5, 7, 35
36	1, 2, 4, 6, 9, 18, 36
37	1, 37
38	1, 2, 19, 38
39	1, 3, 13, 39
40	1, 2, 4, 5, 8, 10, 20, 40

Prime number list

2	179	419	661
3	181	421	673
5	191	431	677
7	193	433	683
11	197	439	691
13	199	443	701
17	211	449	709
19	223	457	719
23	227	461	727
29	229	463	733
31	233	467	739
37	239	479	743
41	241	487	751
43	251	491	757
47	257	499	761
53	263	503	769
59	269	509	773
61	271	521	787
67	277	523	797
71	281	541	809
73	283	547	811
79	293	557	821
83	307	563	823
89	311	569	827
97	313	571	829
101	317	577	839
103	331	587	853
107	337	593	857
109	347	599	859
113	349	601	863
127	353	607	877
131	359	613	881
137	367	617	883
139	373	619	887
149	379	631	907
151	383	641	911
157	389	643	919
163	397	647	929
167	401	653	937
173	409	659	941

famous problems

Throughout the history of mathematics there have been particular problems which have not only been important in themselves but also because of the interest they attracted. Important, that is, for the amount of new mathematics generated as people tried either to solve them or, just as importantly, to prove that they could not be solved. **Fermat's last theorem** is perhaps the most famous of all.

geometrical constructions are accurate diagrams drawn as an answer to a problem, using only a **pair of compasses** and a **straight-edge**. The ancient Greeks, with whom much of our mathematics started, were mainly interested in geometry. To them, most problems were expressed in terms of geometry and their solutions were restricted to being geometrical constructions. So, to divide a given straight line into equal parts, it was not acceptable to do it by measuring, it had to be done by a geometrical construction.

trisection of an angle One of the earliest **geometrical constructions** devised was that for the bisection of an angle – dividing it into two equal parts. Not unreasonably, mathematicians then sought to find a way of trisecting an angle – dividing it into three equal parts. Not until the 19th century was that proved to be impossible.

squaring the circle A very early problem required a square to be made, using a **geometric construction**, that was equal in area to a given circle. This meant finding a length e for the edge of the square so that $e^2 = \pi r^2$ where r is the radius of the circle. This means $e = r\sqrt{\pi}$. For over 2000 years mathematicians struggled with this problem. 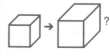 Many close approximations were found, but never an exact solution. In 1882 it was finally proved that it was impossible by any geometrical construction.

duplication of a cube In ancient times, just as now, problems were dressed up in the form of a story. One concerned the temple at Delphi (in Greece) where there was a famous oracle which was often consulted for advice. The story said that, in order to avert a plague, the oracle had required that a new stone altar should be made in the shape of the original one (a cube) but having twice the volume. In modern terms, starting with a cube whose length of edge is e, the cube to be made has to have an edge of length k so that $k^3 = 2e^3$ or $k = e \times \sqrt[3]{2}$. As with the **trisection** problem, at first sight it doesn't look difficult. In two dimensions the equivalent problem is to draw a square which is twice the area of a given square. In that case $k = e \times \sqrt{2}$ and k is easily constructed as shown. Not until the 19th century was it proved that there was no possible **geometrical construction** for $k = e \times \sqrt[3]{2}$.
*It is also known as the **Delian problem**.*

Mascheroni constructions The Italian mathematician Lorenzo Mascheroni (1750–1800) proved that any **geometrical construction** which was possible with compasses and straight-edge could also be done using only a pair of compasses. *However, knowing that it can be done is one thing, finding how to do it is yet another problem.*

cyclotomy is the topic concerned with dividing the circumference of a circle into equal parts, using only **geometrical constructions**. 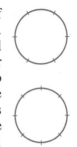 The early Greek mathematicians knew it was possible for all cases where the number of divisions was 2^n, 3 or 5 and all other numbers obtained by multiplying any two of those together. So it was possible for 2, 3, 4, 5, 6, 8, 10, 12, 15, ... divisions. The problem of whether other divisions might be possible was unresolved until **Gauss** (who started on the problem when he was 19) proved that it was possible to construct $2^{2^n} + 1$ divisions provided only that the expression yielded a prime. That added 17; 257 and 65 537 ($n = 2$, 3 and 4) to the list.

four-colour problem In the mid-1800s mathematicians became aware of a map-colouring problem which can be stated simply as: 'For any map that might be drawn on a flat surface, what is the least number of colours that are needed to colour it in such a way that no two countries which touch along a common border have the same colour?' Everyone who attempted it became convinced the answer was 4, but no one could prove it. It was not until 1976 (over 100 years after the problem was first stated) that a proof was found. *That meant it could then be called the four-colour theorem. The drawing on the right shows just how complicated the 'maps' can become, but 4 colours are always enough – provided they are applied correctly.*

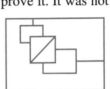

bridges of Königsberg In the 1600s there was a city called Königsberg (it is now called Kaliningrad). The city's buildings were spread over the banks of a river and an island, with all the four principal areas of land connected by 7 bridges. It was said to be a standard challenge to devise a route, starting anywhere, that would cross each bridge once and once only. It was generally believed to be impossible, but no one could prove it could not be done. In 1736 someone gave the problem to **Euler** who proved no answer was possible and, in doing so, started the branch of mathematics now known as **topology**. *It is of interest to consider how the problem would have been changed if there had been six or eight bridges.*

factorising a number is the process by which a whole number is broken up into two (or more) numbers which can be multiplied together to make the original number. Usually it is the **prime factors** which are sought. So 2001 can be factored into $3 \times 23 \times 29$. Finding ways of doing this has fascinated mathematicians for almost as long as there has been arithmetic. Nowadays the problem concerns the finding of faster methods that can be used, with a computer, to find the factors of very big numbers (typically over 100 digits long) in as short a time as possible – to do it in hours or days, rather than in months or years. *This problem has become of practical interest in recent years because of modern encryption methods which use very large numbers having only two prime factors.*

formulas for shapes

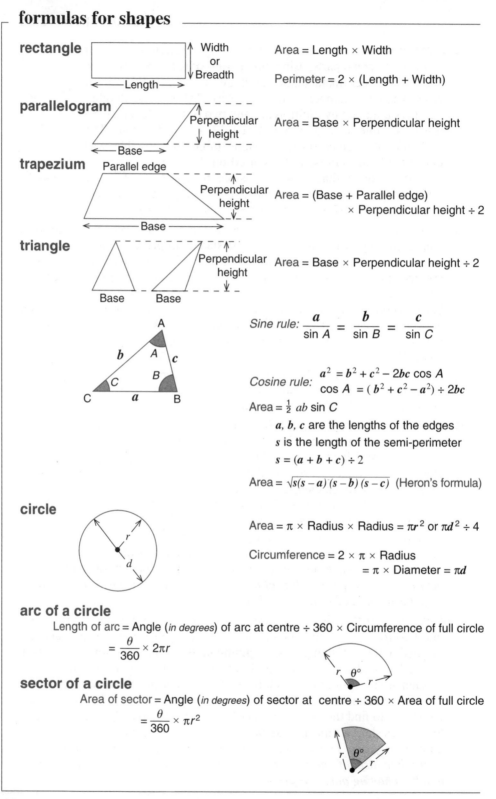

rectangle

Width or Breadth

Length

Area = Length × Width

Perimeter = 2 × (Length + Width)

parallelogram

Perpendicular height

Base

Area = Base × Perpendicular height

trapezium

Parallel edge

Perpendicular height

Base

Area = (Base + Parallel edge)
$\qquad$ × Perpendicular height ÷ 2

triangle

Perpendicular height

Base $\qquad$ Base

Area = Base × Perpendicular height ÷ 2

Sine rule: $\dfrac{a}{\sin A} = \dfrac{b}{\sin B} = \dfrac{c}{\sin C}$

Cosine rule: $\begin{aligned} a^2 &= b^2 + c^2 - 2bc\cos A \\ \cos A &= (b^2 + c^2 - a^2) \div 2bc \end{aligned}$

Area = $\tfrac{1}{2}\,ab\sin C$

a, b, c are the lengths of the edges

s is the length of the semi-perimeter

$s = (a + b + c) \div 2$

Area = $\sqrt{s(s-a)(s-b)(s-c)}$ (Heron's formula)

circle

Area = π × Radius × Radius = πr^2 or $\pi d^2 \div 4$

Circumference = 2 × π × Radius
$\qquad\qquad\qquad = π × Diameter = \pi d$

arc of a circle

Length of arc = Angle (*in degrees*) of arc at centre ÷ 360 × Circumference of full circle

$= \dfrac{\theta}{360} × 2\pi r$

sector of a circle

Area of sector = Angle (*in degrees*) of sector at centre ÷ 360 × Area of full circle

$= \dfrac{\theta}{360} × \pi r^2$

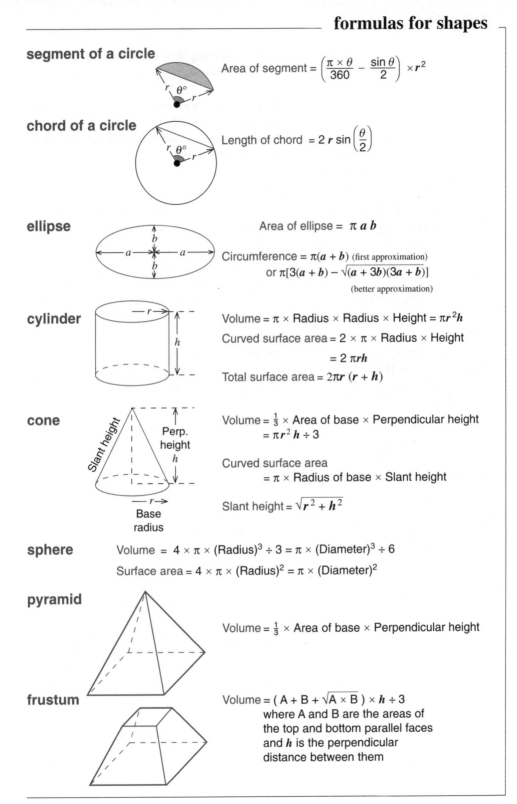

segment of a circle

Area of segment $= \left(\dfrac{\pi \times \theta}{360} - \dfrac{\sin \theta}{2} \right) \times r^2$

chord of a circle

Length of chord $= 2\,r \sin \left(\dfrac{\theta}{2} \right)$

ellipse

Area of ellipse $= \pi\,a\,b$

Circumference $= \pi(a + b)$ (first approximation)

or $\pi[3(a + b) - \sqrt{(a + 3b)(3a + b)}]$

(better approximation)

cylinder

Volume $= \pi \times$ Radius $\times$ Radius $\times$ Height $= \pi r^2 h$

Curved surface area $= 2 \times \pi \times$ Radius $\times$ Height

$= 2\,\pi r h$

Total surface area $= 2\pi r\,(r + h)$

cone

Volume $= \frac{1}{3} \times$ Area of base $\times$ Perpendicular height

$= \pi r^2 h \div 3$

Curved surface area

$= \pi \times$ Radius of base $\times$ Slant height

Slant height $= \sqrt{r^2 + h^2}$

sphere

Volume $= 4 \times \pi \times (\text{Radius})^3 \div 3 = \pi \times (\text{Diameter})^3 \div 6$

Surface area $= 4 \times \pi \times (\text{Radius})^2 = \pi \times (\text{Diameter})^2$

pyramid

Volume $= \frac{1}{3} \times$ Area of base $\times$ Perpendicular height

frustum

Volume $= (\,A + B + \sqrt{A \times B}\,) \times h \div 3$

where A and B are the areas of
the top and bottom parallel faces
and h is the perpendicular
distance between them

fraction A fraction is a measure of how something is to be divided up or shared out. *There are four principal ways of expressing fractions: common, decimal, percentage and ratio.*

common fraction A common fraction is a **fraction** written in the form of two whole numbers, one above the other, separated by a line. The bottom number must not be a 1 or zero. *It represents a division to be done, where the upper number is to be divided by the bottom.*
    Examples: $\frac{1}{2}$ $\frac{3}{4}$ $\frac{2}{3}$ $\frac{99}{150}$ $\frac{-1}{4}$

vulgar fraction ≡ **common fraction**

numerator The numerator is the top number in a **common fraction.**

denominator The denominator is the bottom number in a **common fraction.**
    Example: In the fraction $\frac{8}{9}$, *8 is the numerator, 9 is the denominator*

lowest common denominator or **lcd** The lowest (or least) common denominator of two (or more) fractions is the smallest number into which all of their **denominators** will divide. *It is the* **lcm** *of the denominators.*
    Example: For the fractions $\frac{2}{3}$ $\frac{1}{8}$ $\frac{5}{6}$ *their lcd is 24*

proper fraction A proper fraction is a **common fraction** in which the **numerator** *(= top number)* is smaller than the **denominator** *(= bottom number).*

improper fraction An improper fraction is a **common fraction** in which the **numerator** *(= top number)* is bigger than the **denominator** *(= bottom number).*
    Examples: $\frac{9}{8}$ $\frac{4}{3}$ $\frac{100}{17}$ *are all improper fractions.*

mixed number A mixed number is made up of two parts: a whole number followed by a **proper fraction**.
    Examples: $1\frac{1}{2}$ $5\frac{7}{8}$ $-2\frac{5}{6}$ *are all mixed numbers.*

decimal fraction A decimal fraction is a way of expressing values of fractions less than 1 using the normal **decimal place-value system** extended to the right of the units column so as to give values of $\frac{1}{10}$, $\frac{1}{100}$, etc.
    Example: 0.376 *means* $\frac{3}{10} + \frac{7}{100} + \frac{6}{1000} = \frac{376}{1000}$

decimal point A decimal point is a dot used to show that the values which follow make up a **decimal fraction**. *A comma is used in the metric and SI systems.*

recurring decimal A recurring decimal is a **decimal fraction** which goes on REPEATING itself without end.
    Examples: 0.3333333 ...   *usually written* 0.$\dot{3}$
             0.14285714285714285714 ...   *usually written* 0.$\dot{1}4285\dot{7}$
                where the dot or dots above the number show what is to be repeated; it may be either a single digit or a block of digits.

terminating decimal A terminating decimal is a **decimal fraction** that ends after a definite number of digits have been given.
    Examples: 0.5   0.123   0.67   0.747474

periodic decimal ≡ **recurring decimal**. *Its period is the number of digits which are repeated each time.*   *Example:* 0.$\dot{1}4285\dot{7}$ *has a period of 6*

per cent   indicates a special type of **fraction** in which the value given is a measure of the number of parts in every 100 parts that is to be used.
Example: 25 per cent means 25 in every 100 or $\frac{25}{100}$

% = per cent   *Example: 75% is* 75 *per cent which is* $\frac{75}{100}$

ratio is used to compare the sizes of two (or more) quantities.
Example: Mortar for building a brick wall is made by mixing 2 parts of cement to 7 parts of sand. (The parts may be decided by weight or by volume, just so long as the same units are used.) Then it can be said that the ratio of cement to sand is 2 to 7 which is also written in the form 2:7

equivalent fractions are two, or more, fractions that have the same value but are different in form.
Example: The set of fractions $\frac{3}{4}$ $\frac{6}{8}$ $\frac{63}{84}$ 75% 0.75
all look different, but they all have the same value.

Table of values of some equivalent fractions						
common fraction		decimal fraction	%	common fraction		decimal fraction	%
	$\frac{1}{20}$	0.05	5	$\frac{5}{10}$ $\frac{1}{2}$		0.5	50
$\frac{1}{10}$		0.1	10	$\frac{6}{10}$ $\frac{3}{5}$		0.6	60
$\frac{2}{10}$	$\frac{1}{5}$	0.2	20		$\frac{2}{3}$	0.666...	$66\frac{2}{3}$
	$\frac{1}{4}$	0.25	25	$\frac{7}{10}$		0.7	70
$\frac{3}{10}$		0.3	30		$\frac{3}{4}$	0.75	75
	$\frac{1}{3}$	0.333...	$33\frac{1}{3}$	$\frac{8}{10}$ $\frac{4}{5}$		0.8	80
$\frac{4}{10}$	$\frac{2}{5}$	0.4	40	$\frac{9}{10}$		0.9	90

To change a **common fraction** into a **decimal fraction**, divide the top number by the bottom
   Example: To change $\frac{3}{7}$ *work out* 3 ÷ 7 = 0.428 571 ...

To change a **decimal fraction** into a **percentage,** multiply it by 100
   Example: To change 0.428571 *work out* 0.428 571 × 100% = 42.8571%

To change a **common fraction** into a **percentage,** do both the above in that order.
   Example: To change $\frac{3}{7}$ *work out* 3 ÷ 7 × 100% = 42.8571 ...%

reduced fraction   A reduced fraction is a **common fraction** in its simplest possible form. *To get this, divide both the top and bottom numbers of the fraction by the SAME VALUE until it becomes impossible to do so anymore.*
Example: To reduce $\frac{150}{240}$ *divide both by* 10 *to get* $\frac{15}{24}$,
             then divide both by 3 *to get* $\frac{5}{8}$

algebraic fractions are rather like **common fractions** in their form, but use algebraic expressions for their numerator and/or denominator.
   Examples: $\dfrac{a}{b}$ $\dfrac{x+y}{3x-y}$ $\dfrac{3(x+y)(x-y)}{8(x^2+y)}$

geometry is the study of the properties and relationships of **points**, **lines** and **surfaces** in space.

plane geometry is **geometry** confined to two-dimensional space only.

Euclidean geometry is the **geometry** that keeps within the rules as laid down by Euclid. *It is what is usually meant when the word 'geometry' is used without any other descriptor. It is also the geometry which is most often used in the ordinary, everyday world.*

point A point only indicates a position and has no size. *In a drawing it must have some size in order to be seen, but in any work involving a point its size is ignored. It has no dimensions.*

line A line is the path followed by a **point** when it moves from one position to another so that it has a measurable size (its length) only along that path. *The drawing of a line requires that it has some width in order to be seen, but this size is ignored in all work that would be affected by it. It has only one dimension.*

straight line A straight line is the **line** between two **points** having the least measurable size. *Generally, the word 'line' used by itself means 'straight line' unless the context indicates otherwise.*

linear The word linear is used to indicate an association with a straight line.

line segment A line segment is a piece of a **straight line**. *Strictly speaking, a straight line is fixed by two separate points and goes on indefinitely in both directions so that it cannot be measured. It is only the line segment which is measurable. The single word 'line' is usually taken to mean 'line segment'.*

surface A surface is the two-dimensional outer boundary (or skin) of a three-dimensional object. *It follows next in order after point and line so that now sizes can be measured in two dimensions. A surface is considered to have* NO *thickness.*

plane A plane surface is one where, if ANY two **points** on it are joined by a **straight line,** the line lies entirely on that surface. *More commonly it is known as a flat surface.*

parallel Two (or more) **lines**, which must lie in the same **plane**, are said to be parallel if, no matter how far they are extended in either direction, they are always the same distance apart. *Usually this is applied only to straight lines, but it can be applied to curves that remain a constant distance apart.*

perpendicular Two **straight lines** (or **planes**) are said to be perpendicular to each other if, at their crossing or meeting, a right angle is formed.

orthogonal ≡ **perpendicular**.

collinear Three, or more, **points** are said to be collinear if one **straight line** can be drawn which passes through ALL of them.

opposite angles When two **straight lines** cross each other four angles are made; any pair of these which touch each other only at the crossing point are opposite angles. *Opposite angles are equal in size and also known as* **vertically opposite angles**.

adjacent angles When two **straight lines** cross each other four angles are made; any pair of these which touch each other along a line are adjacent angles. *Adjacent angles add up to 180°.*

transversal A transversal is a straight line that cuts across other straight lines. *The other lines are usually parallel.*

alternate angles When a **transversal** cuts two **parallel** lines alternate angles are any pair of angles that lie on OPPOSITE sides of the transversal and on OPPOSITE relative sides of the parallel lines. *Alternate angles are equal in size. They are also known as* **Z-angles**.

corresponding angles When a **transversal** cuts two **parallel** lines corresponding angles are any pair of angles which lie on the SAME sides of the transversal and on the SAME relative sides of the parallel lines. *Corresponding angles are equal in size.*

similar Geometrical figures are said to be similar if they are the SAME in shape but DIFFERENT in size. *One shape is an enlargement of another. Corresponding angles in each shape will be the same size. All these triangles are similar:*

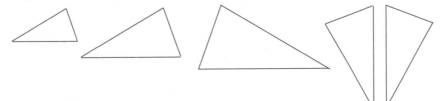

congruent Geometrical figures are said to be congruent if they are the SAME in shape AND size. *One shape can be fitted exactly over the other, being turned around and/or over as necessary. Such shapes are also described as being identical or equal. These four triangles are all congruent:*

vertical A vertical line at any point on the Earth is the **straight line** which would join that point to the centre of the Earth. *Usually, in drawings, a vertical line means one which goes in the top to bottom direction of the page.*

horizontal A horizontal line at any point on the Earth is a **straight line** that lies at right angles to the **vertical** at that point. *It is often described as 'level'. Usually, in drawings, a horizontal line means one which goes across the page.*

graph   A graph is a diagram showing EITHER the relationship between some variable quantities OR the connections that exist between a set of points (as in **topology**). *The word 'graph' on its own usually means the first kind, where the relationship is shown by means of points plotted on a coordinate system, and is the type covered in this section. Statistical graphs are also of the first kind but are usually named by type, such as bar chart, pictogram, etc. For most work the relationship concerns only two quantities.*

quadrants   The 2 axes of a coordinate system divide the plane into 4 separate sections known as quadrants. These are identified as the first, second, third and fourth quadrants in the way shown on the right.

second 2nd	first 1st
third 3rd	fourth 4th

linear graph   A linear graph is a **graph** in which all the points representing the relationship between the quantities lie on a straight line.

intercept   The intercept of a **graph** is the point at which it cuts across an axis. *For linear graphs this word is usually reserved for the point at which the line cuts the y-axis. Example: In the linear graph shown on the right the intercept is ⁻2*

gradient   The gradient of a line drawn on a **graph** is a measure of its slope relative to the *x*-axis. *This is expressed as the ratio of its vertical change to its horizontal change, both changes being measured on the scales of their respective axes. Example: In the diagram the gradient is given by a ÷ b and is positive since y INCREASES as x increases.*

negative gradient   A negative gradient is a **gradient** which shows that *y* DECREASES as *x* increases. *It is measured as for gradient (above) but has a negative sign in front of the value.*

$y = mx + c$ is the equation of a straight line which has a **gradient** of value *m* and an **intercept** (on the *y*-axis) at *c*. *Example: y = 3x – 4 has a gradient of 3 and an intercept of ⁻4*

trend line   A trend line is the single line that best represents the general direction of a set of points. *This is especially useful when some observed data has been plotted that does not lie on a straight line, but estimates are required to be made based on that data. It is also known as a* **line of best fit**.

interpolation   An interpolation is an estimation of the likely value of an unknown piece of data, falling WITHIN the range of some known data and based on the evidence provided by that known data. *This usually involves using the trend line. Example: In the graph on the right the values of the known data are shown by black dots, and the interpolated value by a red dot.*

extrapolation An extrapolation is an estimation of the likely value of an unknown piece of data, falling OUTSIDE the range of some known data and based on the evidence provided by that known data. *This usually involves using the trend line. Example: In the graph on the right the values of the known data are shown by black dots, and the extrapolated value by a red dot.*

quadratic graph A quadratic graph is a **graph** in which the relationship between the variables is given by a **quadratic equation**. *Its shape is that of a parabola. Examples:*

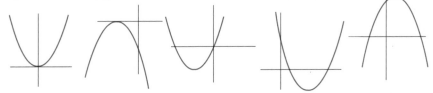

roots of a quadratic The two **roots** of a **quadratic expression**, when it is drawn as a graph, are indicated by the points at which the line of the graph crosses the *x*-axis. *If the line crosses the x-axis, the two x-values at those crossings are the roots of that quadratic. If the line touches, but does not cross, the x-axis, the two roots are equal. If the line does not cross or touch the x-axis, the roots are complex numbers. Example: The graph on the right of the quadratic $x^2 - x - 2$ has its roots at $^-1$ and 2*

cubic A cubic graph is a **graph** in which the relationship between the variables involves an expression of **degree** 3
Example: The graph of $y = x^3 - 4x^2 - 15x + 18$ is shown on the right. Its roots are $^-3$, 1 and 6

inverse square An inverse square relationship is $y = k/x^2$, where *k* can be any number. *A graph showing this relationship has a general shape like that shown on the right. If k is negative, then it will be rotated into the 4th quadrant.*

exponential An exponential graph is a **graph** in which the relationship between the variables involves an expression in which one of the variables appears as an **exponent** or **index**. *The word is often used in connection with 'exponential growth' to indicate that its rate of change is always increasing. Example: $y = 2^x$ is an exponential relationship.*

inequalities An inequality is shown by an expression such as $y < 3 - x$, meaning that *y* can take any value which is LESS THAN that of $3 - x$. *This is shown on a graph by drawing the line of $y = 3 - x$ and shading the region where y is always less than that.*

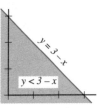

information technology is the term used to include a whole range of electronic devices and techniques used in collecting, storing, retrieving, processing, presenting and transmitting data. *Computer systems are usually a major item in all of these activities, but are not exclusively so.*

computer A computer is a machine which will perform any operations that can be expressed in terms of logic and arithmetic. *Its power (and its usefulness) lies in the speed at which it works. Modern computers are capable of carrying out over 300 million instructions in every second.*

program A program is the set of instructions, written in a particular language which a **computer** can understand, to tell the computer exactly how the required operations are to be done.

programming language A programming language consists of a set of precisely defined rules, and the exact way in which they are to be written, so as to be understood by a computer. *There are several different languages, each having been written for some specific purpose.*

BASIC stands for Beginners All-purpose Symbolic Instruction Code. This is a **programming language** originally developed for educational purposes that is being continually extended for use in commercial applications.

C and **C++** are both **programming languages**, C++ being an extension of C. These are now the languages most widely used by professional programmers.

LOGO is a **programming language** developed for educational use.

RAM stands for **Random Access Memory**. This is an electronic store for information which can be written to, or read by, the computer system. The information is LOST when the power is switched off.

ROM stands for **Read Only Memory**. This is an electronic store for information which can be read, but NOT written to, by the computer system. The information is NOT LOST when the power is switched off.

bit A bit is the smallest piece of information that can be stored or transmitted electronically. It can be represented by a 0 or a 1.

byte A byte is a number of **bits** grouped together to make up a single unit of information. Common usage has 1 byte = 8 bits, so that a byte could be 01001110 which is the binary representation of the number 78.

Kb stands for **Kilobyte**, which is 2^{10} or 1024 **bytes**.

Mb stands for **Megabyte**, which is 2^{20} or 1048 576 **bytes**.

Gb stands for **Gigabyte**, which is 2^{30} or 1073 741 824 **bytes**.

database A database is a computer program which allows information to be stored in an organised way so that each separate item can be easily found.

record A record is one set of information in a **database** which is made up of several separate pieces of data linked together by a common theme. *For example, all the information about one person could form a record.*

field A field is one specific piece of information in a **database**, and a collection of fields on a common theme form a **record**.

spreadsheet A spreadsheet is a computer program which presents the user with a large number of cells (like squared paper) each of which can hold a piece of data, and these cells can be linked together in various ways so that a change of a piece of data in one cell immediately produces a change in all those cells to which it is connected. *The data is usually numbers, but does not have to be. The example below shows part of a spreadsheet in action.*

The top left-hand corner of a spreadsheet looks like this. Cells are identified using the letters and numbers to get A1, C2, M15 etc.

	A	B	C	D
1				
2				
3				

The data 7, 6, 4, 5, 3, 8 has been put in, and Cells D 1 to 3 hold totals of all cells to their left. A3 to C3 hold totals of cells above them.

	A	B	C	D
1	7	6	4	17
2	5	3	8	16
3	12	9	12	33

Changing the value in A1 from 7 to 18 immediately causes D1, A3 and D4 to change as they recalculate their own values.

	A	B	C	D
1	18	6	4	28
2	5	3	8	16
3	23	9	12	44

Modern spreadsheets can have over 4 million cells and hundreds of different formulas to link them. They are also capable of presenting the data in a wide variety of ways.

calculator The word calculator usually refers to a small hand-held device that allows numbers to be entered, mathematical operations carried out (electronically), and the results displayed on a very small screen. For a basic model the operations are usually restricted to $+-\times\div\sqrt{}$ and the display shows eight digits. There is sometimes a 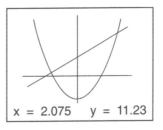 memory facility, but all data is lost when the calculator is switched off.

scientific calculator A scientific calculator is a **calculator** that can carry out many more operations (in trigonometry, statistics, etc.). It has a slightly bigger display so that it can show large numbers in **scientific notation.**

programmable calculator A programmable calculator is a **scientific calculator** in which small programs can be keyed in, so it serves as a simple **computer**. It also has a memory which does not lose its contents when switched off.

graphics calculator A graphics calculator is a **programmable calculator** that has a larger display screen, so that graphs, text and numbers can be shown. It has a larger memory space and no data is lost when the power is switched off.

With each new model that is brought out, more facilities are included, so that graphics calculators are becoming more powerful and more like miniature computers.
The diagram shows the display when the line $y = 3x + 5$ and the parabola $y = 4x^2 - 6$ are drawn on the same axes. It also gives the coordinates of one of the intersections.

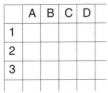

compass and **compasses** are shortened ways of saying a **pair of compasses** which is an instrument consisting of two legs, hinged together at one common end, with the other ends of the legs having, respectively, a sharp point and a drawing implement such as a pen or pencil. *Their principal use is for drawing circles or parts of a circle.*

beam or **trammel compasses** are **compasses** in which the two legs are not hinged together but are held on a single horizontal bar, along which they can be moved and then locked into position. *They serve the same purpose as a pair of compasses, but are more suitable for larger circles because the drawing implement is always perpendicular to the surface over which it moves.*

bow compasses are small **compasses** which are intended for use in drawing small circles. *They usually have a device to allow very fine adjustments to be made. It is their small size which makes them so easy to manipulate, and circles can be drawn with a simple twisting action between finger and thumb.*

dividers or a **pair of dividers** are similar in appearance to **compasses** but in these, both legs have a sharp point at the end. *They are used for measuring, or marking off, the distance between two points on a drawing, a ruler or a scale.*

scaled ruler In maps and drawings which are drawn to **scale** it is usual to provide a scaled ruler on which lengths taken from the drawing can be measured as actual sizes. *The example below shows a scaled ruler that would be suitable for a drawing made to a scale of 1:50. In this case the actual distance represented by the length of the black line would be 3.8 metres.*

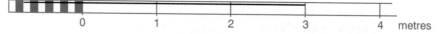

0	1	2	3	4 metres

diagonal scales are used to increase the accuracy with which readings can be taken from a **scaled ruler** by giving markings of one-hundredths of the unit being used. *In the drawing it can be seen how the diagonal lines on the left-hand end divide each tenth into ten equal parts. The length between the two black dots is 1.26 units.*

protractor A protractor is an instrument made as a flat shape (usually a circle or semicircle of clear plastic) with a set of graduated marks, and is used for measuring angles. *Shown here is a common type with dual scales for measuring in either direction.*

template A template is a thin sheet of some stout material (usually plastic) which is cut into a particular shape, or has shapes cut out of it, to provide outlines which can be drawn firmly and quickly. *The example below shows a template that might be used to draw some shapes, symbols and letters. It is sometimes referred to as a* **stencil***.*

French curves are **templates** which are formed from a variety of curves to provide a firm edge and allow a good quality curved line to be drawn. *The drawing on the right shows one French curve from a commercially produced set of three. In use the curve which is needed to be drawn is first sketched lightly in freehand to get a close idea of its shape, and the French curves are then used to drawn the final line, a bit at a time, by matching it in sections. They are mainly used by technical illustrators.*

set squares are **templates** cut in the shape of a right-angled triangle whose two other angles are either both 45° or 30° and 60°. *Adjustable set squares are made with the hypotenuse hinged at one end so that it can be adjusted to form different angles.*

T-square A T-square is an instrument in the shape of the capital letter T which is used on a drawing board for drawing parallel lines. *It is rarely used nowadays by professionals, having been replaced by drawing-boards which incorporate a parallel-motion mechanism. The drawing shows a T-square in place on a drawing-board with some lines drawn.*

straight-edge A straight-edge is a plain unmarked flat bar (usually made of wood, plastic or metal) having at least one of its edges accurately finished so that it can be used to draw a **straight line**. *It has no markings and so cannot be used to make measurements.*

ruler A ruler is a **straight-edge** having graduated markings along at least one of its edges so that lengths (of a straight line) can be measured. *It is sometimes referred to as a* **rule***.*

parallel ruler A parallel ruler (or rulers) is an instrument for drawing lines parallel to an existing line or direction. *One way this is done is by building a pair of rollers into a* **straight-edge** *which allow it to be moved over a flat surface while always staying parallel to the line against which it was originally placed. This type is usually made of metal and of sturdy construction, so that its weight helps to ensure that it rolls over the surface without slipping. It is much used by navigators when working on charts.*

instruments and mechanisms

plumb-line A plumb-line is a length of cord with a weight (usually made of lead) fastened at one end so that when the weight is allowed to hang freely, and is still, the cord will give a line which is **vertical** to the Earth's surface at that place.

level A level is an instrument used to make sure that a line, or the surface of an object, is parallel to the **horizontal** plane of the Earth's surface at that place. *The most commonly seen example of this intrument is the spirit-level in which the movement of an air-bubble inside a curved glass tube will show when the instrument is horizontal.*

rangefinder A rangefinder is an instrument which measures the distance from the observer to some distant object. *The simplest types use two lenses mounted one at each end of a base-line of known length, and the amount these lenses have to be turned to view the object allows a calculation to be made (automatically) of the distance of the object. More sophisticated and more accurate rangefinders work by calculating the distance from the time needed for some form of electromagnetic beam to travel to the object and back again.*

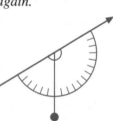

clinometer A clinometer is a hand-held instrument used to measure the **angle of elevation** of some point, in relation to the **horizontal** plane of the observer. *It has either a built-in* **plumb-line** *or* **level**, *and the angle through which the instrument has to be tilted out of the horizontal so as to be aligned with the point being looked at is shown on a suitable scale.*

cross-staff A cross-staff is a simple, easily made, instrument to measure the **subtended angle** of two distant points at the observer's position. *It is an old instrument which was used by astronomers from about 500 BC onwards. It was adopted for sea-going navigation in the early 1500s. There are two parts to the instrument: the main shaft* **EF** *and the cross-bar* **AB** *which is free to slide along* **EF**. *The observer's eye is put close to* **E** *while looking at the two distant points being considered. Cross-bar* **AB** *is then slid along* **EF** *until one of the distant points is coincident with* **A** *and the other is coincident with* **B**. *The angle subtended by those two points at* **E** *can then be read off where* **AB** *cuts across a scale marked on* **EF**.

sextant A sextant is a hand-held instrument used to measure the **subtended angle** at the oberver's position between two distant points. *It uses a mirror so that the observer can see both points at the same time, and can move the mirror so as to bring them into line; the amount of movement needed to do this gives the size of the subtended angle. It is used mostly by navigators to find the elevation above the horizon (known as the altitude) of the Sun, the Moon or a star. It was invented in the 1730s, and the name is derived from the fact that its calibrated scale is an arc which is one-sixth of a circle.*

sine bar A sine bar is a steel bar, made to a high degree of accuracy, which allows an angle to be set up by measuring lengths rather than by measuring the actual angle. *The bar has two identical cylindrical pins fitted into its long surface, one at each end, with a known distance between their centres (typically 100, 200 or 300 mm). Steel blocks of known thickness can be built up on which these pins rest, and so incline the bar at an angle to the base surface. Since all relevant measurements are known, the sine of the angle can be calculated, and thus the angle itself. In practice, it is usually the angle which has to be set up, and so the heights of the blocks have to be calculated.*

spherometer A spherometer is an instrument for measuring the radius of curvature of a surface which is assumed to be part of a sphere. *It consists of a base-plate on which 3 legs of equal length are fixed perpendicular to the base-plate. The points of these legs define an equilateral triangle. When placed on a curved surface, a screw in the middle of the base-plate is adjusted until it just touches the surface. The radius of curvature can then be calculated, or read off on a suitably scaled dial.*

pantograph A pantograph is an instrument used to copy drawings while making them either smaller or bigger but always keeping them **similar** to the original. *In the outline drawing of a pantograph below, the 4 links AQ, AF, CF, BC are joined together at A, B, C, and F, but are free to swivel at those points. Point F is held still. AF = BC and AB = FC. If the lines of the diagram to be copied are followed by point P, then point Q will trace out an* **enlargement** *of the original diagram. A reduction can be made by changing over the roles of P and Q. The scale of the enlargement or reduction can be changed by altering the relative positions of P and Q along their respective links.*

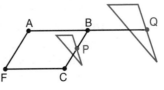

planimeter A planimeter is an instrument used for measuring the area of any shape on a plane drawing. *The simplest type is shown in the diagram. PI and IT are two arms which are free to move. P is a fixed point. T is used to trace around the perimeter of the drawing. The relative movements of the two arms as T moves is recorded by a unit fixed on the joint at I, known as the integrator, which then displays the size of the (shaded) area enclosed by T's movements.*

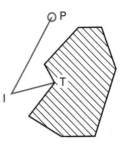

Peaucellier's straight line mechanism is a system of links devised (in 1864) to enable a rotary motion to be changed into movement along a straight line. *It uses a total of 7 links: LR and LS which are equal in length; PR, PS, QR and QS all equal in length; and OP which is equal in length to distance OL. O and L are fixed in position. As P turns in a circle about O, Q moves up and down in a straight line.*

kinematics is the branch of mathematics dealing with the motion of objects and considering only their movement in relation to space and time, disregarding the effects of their mass or any forces acting upon them.

speed The speed of a moving object is a measure of the distance travelled by the object in a unit period of time. *Any suitable units may be used for the distance (metres, feet, miles, kilometres, etc.) and the time (hours, minutes, seconds, etc.).*

average speed When an object moves through some distance its **speed** may vary as it travels, but its average speed is found by considering only the total time taken for the move and the total distance moved. *The three equations used in working with average speed are:*

Speed = Distance ÷ Time	Time = Distance ÷ Speed
Distance = Speed × Time	

Example: A car is driven from Aberdeen to Bristol, a distance of 780 km, and the journey takes 20 hours. What is the average speed for the journey? The average speed is 780 ÷ 20 = 39 km per hour.

velocity The velocity of a moving object is given by stating both its **speed** and the direction in which the object is moving. *It is a* **vector** *quantity. If the velocity is stated without reference to any direction, it must be assumed that the object is travelling in a straight line and that the overall direction of that line does not matter. The word 'velocity' is very commonly used as being the same as 'speed'. The SI unit of velocity is metres per second, abbreviated to m/s or m s^{-1}.*

acceleration The acceleration of a moving object is a measure of how its **velocity** is changing in relation to time. *It is a vector quantity but its direction is often ignored and it is applied only to the speed component of the velocity. It may be positive (for speeding up) or negative (for slowing down). The SI unit of acceleration is metres per second per second, abbreviated to m/s^2 or m s^{-2}. Example: Starting from rest (= zero velocity) an object is given a steady acceleration of 3 m s^{-2}. What is its velocity after 10 seconds? In this case the velocity will increase by 3 m s^{-1} after every second of its travel. So its velocity after 1 second is 3 m s^{-1}, after 2 seconds is 6 m s^{-1}, after 3 seconds is 9 m s^{-1}, ..., after 10 seconds is 30 m s^{-1}.*

deceleration If the **velocity** of a moving object is decreasing, then its **acceleration** is negative and is often described as deceleration.

retardation ≡ deceleration

constant '...' When the property named in '...' does not change during the period of time being considered, that property is described as constant. *So we have constant speed, constant velocity, constant acceleration.*

uniform '...' ≡ constant '...'

displacement The displacement of an object in motion is the distance and direction from its starting position to its finishing position. *It is a* **vector**.

distance–time graph A distance–time graph is a **graph** that shows the relationship between the distance moved by an object in relation to time. *The* **gradient** *of a line drawn on the distance–time graph is a measure of* **velocity**. *If the line of the relationship is curved, the particular velocity at some moment in time can only be found by drawing a* **tangent to the curve** *at that point and measuring its gradient. It is also known as a* **travel graph**.

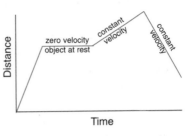

velocity–time graph A velocity–time graph is a **graph** that shows the relationship between the velocity of an object in relation to time. *The* **gradient** *of a line drawn on the graph is a measure of* **acceleration**. *If the line of the relationship is curved, the particular acceleration at some moment in time can only be found by drawing a* **tangent to the curve** *at that point and measuring its gradient. The* **area under the curve** *(or the line of relationship) between two* **ordinates** *drawn from the time scale measures the distance travelled in that time interval.*

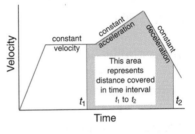

compound measures are those measures that require more than one unit to give their value. *Units such as metres, seconds, feet etc. are used for single measures. One of the most commonly used compound measures is speed (in units of metres/second, miles/hour, etc.). Other examples of compound measures are density, work, power, fuel consumption.*

To change compound units of speed from one in the left-hand column into one of those on the right, **multiply** by the number given in the table.

To change ↴ into →	m.p.h	km.p.h	m s⁻¹	ft/second
miles per hour (m.p.h)	1	1.6093	0.44704	1.4667
kilometres per hour (km.p.h)	0.62137	1	0.27778	0.91134
metres per second (m s⁻¹)	2.2369	3.6	1	3.2808
feet per second (ft/second)	0.68182	1.0973	0.3048	1

equations of uniform motion There are five equations giving the relationships between the five variables controlling movement under conditions of uniform motion:

$v = u + at$		u is the initial velocity
$s = \frac{1}{2}(u + v)\,t$		v is the final velocity
$s = ut + \frac{1}{2}at^2$	where	s is the distance moved
$s = vt - \frac{1}{2}at^2$		a is the acceleration
$v^2 = u^2 + 2as$		t is the time taken

logic The logic of a system is the whole structure of rules that must be used for any reasoning within that system. *Most of mathematics is based upon a well-understood structure of rules and is considered to be highly logical. It is always necessary to state, or otherwise have it understood, what rules are being used before any logic can be applied.*

statement A statement made within a **logical** system is a form of words (or symbols) which carries information. *Within mathematics nearly everything is written in the form of statements.*
Examples: *The length of the radius is 4 cm.* $3x + 2 = 7$

argument An argument is a set of one or more **statements** which uses the **logic** of the system to show how one particular statement is arrived at.

true Within a system, a **statement** is said to be true when it is a known fact, or follows from some other true statement by means of a **valid argument**, or is considered to be **self-evident**.

false Within a system, a **statement** is said to be false when it is contrary to a statement known to be **true**.

undecidable Within a system, a **statement** is said to be undecidable when it cannot be shown to be either **true** or **false**.

assumption An assumption is a **statement** (true or false) which is to be taken as **true** for the purpose of the **argument** which follows.

premise $\equiv$ **assumption**

self-evident A **statement** is described as self-evident when it is thought that no reasoning is necessary to demonstrate that the statement is **true**. *This is often used to describe the most basic ideas of a system which are generally 'known' but are impossible to define independently of the system.*
Example: *The statement 'Any two things which are each equal to a third thing must be equal to each other' could be seen as being self-evident.*

intuitive Understanding (of a statement or a piece of knowledge) is described as intuitive when it is, or can be, reached without support of any **argument**.

axiom An axiom is a **statement** which is assumed to be **true**, and is used as a basis for developing a system. *Any system of logic starts by saying clearly what axioms it uses.*

proposition A proposition is a **statement** whose correctness (or otherwise) is to be shown by the use of an **argument**. *It most often serves as an introduction by saying, in effect, what the argument is going to show.*

valid A valid **proof** (or **statement**) is one in which all the **arguments** leading up to it are correct within the logic of the system being used.

invalid An invalid **proof** (or **statement**) is one which is NOT **valid**.

counter-example A counter-example to a **statement** is a particular instance of where that statement is not **true**. *This makes the statement invalid.*
It only requires ONE counter-example to make a statement invalid.
Example: *'All prime numbers are odd': a counter-example is 2*

proof A proof is a sequence of **statements** (made up of **axioms**, **assumptions** and **arguments**) leading to the establishment of the **truth** of one final statement.

direct proof A direct proof is a **proof** in which all the **assumptions** used are **true** and all the **arguments** are **valid**.
Example: To prove the proposition that adding two odd numbers makes an even number. Any odd number is of the form $2n + 1$; (n is a whole number).
$$Adding\ two\ of\ this\ form\ produces\ (2n + 1) + (2m + 1)$$
$$= 2(n + m) + 2 = 2(n + m + 1),\ which\ is\ clearly\ even.$$

indirect proof An indirect proof is a **proof** in which one **assumption** is made. Then, using **valid arguments**, a **statement** is arrived at which is clearly false; so the original assumption must have been false. *This can only be used in a system in which statements must be either true or false, so that proving the first assumption is false allows only one possibility for its alternative form – which must be the correct one.*
Example: To prove $\sqrt{2}$ is irrational. First assume that it is rational.
Then $\sqrt{2} = \frac{a}{b}$, where a, b are whole numbers with no common factors.
This leads to $a^2 = 2b^2$, and a^2 must be even and so must a.
Put $a = 2c$ then $\sqrt{2} = \frac{2c}{b}$ and $2c^2 = b^2$ and b must be even.
But a,b had no common factors so both cannot be even.
The assumption must be false and $\sqrt{2}$ is NOT rational.
$\sqrt{2}$ must be irrational.

proof by contradiction ≡ **indirect proof**

reductio ad absurdum ≡ **indirect proof**

proof by exhaustion A proof by exhaustion is a **proof** which is established by working through EVERY possible case and finding no contradictions. *Usually such a proof is only possible if the proposition to be proved has some restrictions placed upon it.*
*Example: The statement that 'Between every pair of square numbers there is at least one prime number' would be impossible to prove by looking at every possibility. However, by writing it as 'Between every pair of square numbers less than 1000 there is at least one prime number' it can be proved by exhaustion – looking at every case. This might then be considered as enough evidence to make it a **conjecture** about all numbers.*

proof by induction A proof by induction is a **proof** which shows that IF one particular case is **true** then so is the next one; it also shows that one particular case IS true. *From those two actions it must follow that ALL cases are true.*

visual proof A visual proof is a **proof** in which the **statements** are presented in the form of diagrams.
Example: To prove the proposition that adding two odd numbers makes an even number. Any odd number can be shown as
Adding two odd numbers is shown and clearly makes an even number.

'look-see' proof ≡ **visual proof**

logic (practice)

conjecture A conjecture is a **statement** which, although much evidence can be found to support it, has not been proved to be either **true** or **false**.

hypothesis A hypothesis is a **statement** which is usually thought to be true, and serves as a starting-point in looking for **arguments** (or evidence) to support it. *This word is mostly used in statistics.*

theorem A theorem is a **statement** which has been **proved** to be **true**.

lemma A lemma is a **theorem** which is used in the **proof** of another theorem. *Usually a lemma is of no great importance in itself, but it is a useful way of simplifying the proof of the final theorem by reducing its length.*

corollary A corollary follows after a **theorem** and is a **proposition** which must be **true** because of that theorem.
Example: It can be proved that the three interior angles of a triangle add up to 180° (a theorem). A corollary of this is that the exterior angle at one vertex must equal the sum of the interior angles of the other two vertices.

$$X = B + C$$

converse The converse of a **theorem** (or statement) is formed by taking the conclusion as the starting-point and having the starting-point as the conclusion. *Though any theorem can be re-formed in this way, the result may or may not be true and it needs its own proof.*
Example: One theorem states that if a triangle has two edges of equal length then the angles opposite to those edges are also equal in size. The converse is that if a triangle has two angles of equal size, then the edges opposite to those angles must be equal in length – and that can also be proved.

contrapositive The contrapositive of a **statement** is formed by taking the conclusion as the starting-point and the starting-point as the conclusion and then changing the sense of each (from positive to negative and vice versa). *If the original statement was true, then the contrapositive must also be true.*
Example: The statement 'If a number IS even, it CAN be divided by two' has the contrapositive 'If a number CANNOT be divided by two, then it is NOT even'.

necessary condition A necessary condition for a **statement Q** to be **true** is another statement **P** which MUST be true whenever statement **Q** is true; then statement **P** is said to be a necessary condition. *When P is true then Q may be true or false, but when P is false, then Q must also be false.*
Example: A necessary condition for the statement (Q) 'x is divisible by 6' is statement (P) 'x is even', but condition (P) by itself allows values such as 2, 4, 8 etc. which are clearly not divisible by 6

sufficient condition A sufficient condition for a **statement Q** to be **true** is another statement **P** which, when **P** is true, guarantees that statement **Q** MUST also be true. *When statement P is false then statement Q may be true or false.*
Example: A sufficient condition for the statement (Q) 'x is divisible by 6' is statement (P) 'x is divisible by 12', but condition (P) by itself excludes values such as 6, 18, 30, etc. which are also divisible by 6

necessary and sufficient condition A necessary and sufficient condition for a **statement** to be **true** is a second statement such that BOTH the first and the second statement MUST be true at the same time. *Both statements will be false together as well, but it cannot be that one is true and one is false. Example: A necessary and sufficient condition for the statement 'x is divisible by 6' is that 'x is even and divisible by 3'.*

paradox A paradox is a **valid statement** which is self-contradictory or appears to be wrong. *Paradoxes are important to the development of logic systems. Example: 'The barber shaves all the men in this village who do not shave themselves' seems a reasonably clear statement. However, given that the barber is a man and lives in that village, who shaves the barber?*

fallacy A fallacy is an **argument** which seems to be correct but which contains at least one error and, as a consequence, produces a final statement which is clearly wrong. *Though it is clear that the result is wrong, the error in the argument is usually (and ought to be) difficult to find.*
Example: Let $x = y$: Then $x^2 = xy$ and $x^2 - y^2 = xy - y^2$
This gives $(x + y)(x - y) = y(x - y)$ so that dividing both sides by $(x - y)$
leaves $x + y = y$

From this result, putting $x = y = 1$ means $2 = 1$
Or, subtracting y from both sides means $x (= any number) = 0$
The error is in dividing by $(x - y)$ which is zero.

symbols The conventional way of representing statements is by using capital letters. *The letters most often used are P and Q.*
Example: P could represent 'x is a prime number greater than 2'

P ⇒ Q where P and Q are statements, is a symbolic way of saying 'P implies Q'; OR 'when P is true then so is Q'; OR 'P is a **sufficient condition** for Q'. *It also means that Q is a necessary condition for P.*
Example: If P represents 'x is a prime number greater than 2', and Q represents 'x is an odd number', then $P \Rightarrow Q$.

P ⇐ Q where P and Q are statements, is a symbolic way of saying 'P is implied by Q'; OR 'when Q is true, then so also is P'; OR 'P is a **necessary condition** for Q'. *It also means that Q is a sufficient condition for P.*
Example: If P is 'x is divisible by 5', and Q is 'x is divisible by 10', then $P \Leftarrow Q$; which is a symbolic way of stating that a number which is divisible by 10 is also divisible by 5; or of saying that it is necessary (but not sufficient) that a number is divisible by 5 if it is to be divisible by 10.

P ⇔ Q where P and Q are statements, is a symbolic way of saying that P and Q must both be true (or false) together; OR 'P implies and is implied by Q'; OR 'P is a **necessary and sufficient condition** for Q'.
Example: If P is 'x is divisible by 6', and Q is 'x is divisible by 2 and 3', then $P \Rightarrow Q$ and $Q \Rightarrow P$ so $P \Leftrightarrow Q$.

iff is a short way of writing 'if and only if' and is equivalent to ⇔.
Example: In the previous example the final statement could be 'P iff Q'.

mathematicians of earlier times

The numbers given for each mathematician refer to the places marked on the map opposite showing where that person worked.

Thales ①(c.624–548 BC) is the first mathematician known to us by name. He was a wealthy Greek who travelled widely and worked on many subjects including mathematics, astronomy and philosophy. He appears to have been the first to produce theorems which were supported by logical reasoning rather than experiment. Among other things he showed how it was possible to work out the height of a pyramid from the length of its shadow – using a stick placed vertically in the ground, and a calculation based on shadow lengths and **similar** triangles.

Pythagoras ②(c.572–495 BC) was a Greek mathematician who established a 'school' in what is now Southern Italy, although at that time it was under Greek rule. It is after him that **Pythagoras' theorem** is named. It was a basic belief of the Pythagoreans (as those who worked with him were known) that everything could be explained by numbers – by that they meant whole numbers – and it came as a shock when they discovered, and proved, the existence of **irrational numbers**. They also discovered **deficient, perfect,** and **abundant numbers; amicable pairs** and **polygon numbers;** as well as various relationships between the notes of a musical scale.

Euclid ③(c.365–300 BC) was a Greek mathematician who founded a school of mathematics at Alexandria. He is most famous for writing the longest-lasting textbook ever on mathematics. His book was called *Elements* and was a gathering together of all the important ideas and theorems concerning geometry known at that time. It was the start (and has continued to be the basis) of what is now known as **Euclidean geometry**. As well as gathering together the works of others, he also proved some mathematical theorems himself. Among these were Pythagoras' theorem, the existence of irrational numbers and, most famously, that the number of primes was infinite. He invented what is known as **Euclid's algorithm**.

Archimedes ④(c.287–212 BC) was a Greek mathematician who worked in what is now called Sicily, though at that time it was under Greek rule. He is generally recognised as one of the all-time great mathematicians as his work was so much in advance of anything done before. He was the first to calculate the value of π to three decimal places. He invented a way of calculating the volumes of pyramids, cones and other shapes by a method that was not properly appreciated for more than a 1000 years. He devised a system for writing very large numbers that can be seen as a forerunner of our present system of **index notation**, and his name is attached to various mathematical ideas such as **Archimedean solids** and **Archimedes' spiral**. He was a very practical man and also has a claim to fame as an engineer and a scientist.

Eratosthenes ⑤(c.276–196 BC) was a Greek with interests in many subjects including mathematics and astronomy. Because of his wide-ranging knowledge he was invited by Ptolemy III to superintend the Great Library of Alexandria. He devised the method of finding prime numbers known as **Eratosthenes' sieve** and calculated the circumference of the Earth, with surprising accuracy, as 40 000 km.

Heron ⑥(c.65–125 AD) was a mathematician who worked in Alexandria and was the first to develop ideas in what is generally known as 'applied mathematics' or 'mechanics'. He proved and published **Heron's formula** for the area of a triangle, though it was probably known about before then.

Claudius Ptolemy ⑦(c.90–168 AD) was an astronomer and geometer who worked in Alexandria. His great work was the *Almagest* which gave details of the movements of the stars and planets and explained how the universe moved around the Earth. This idea was firmly held for over a thousand years. He also calculated the value of π to four decimal places, and the square root of 3 to five decimal places.

Diophantus ⑧(c.200–284 AD) was a Greek mathematician who worked at Alexandria. He was the first to solve problems by what we would regard as algebraic (rather than geometric) methods. He also originated **Diophantine equations**.

Hypatia ⑨(370–415 AD) is the first woman known to have been a mathematician. She headed her own school in Alexandria and many important men of that time attended. The fact that she was an important pagan eventually led to her brutal murder by Christian fanatics.

al'Khwarizmi ⑩(c.780–850 AD) was an Arabian mathematician who worked in Baghdad. His most important book *Ilm al-jabr wa'l muqabalah* introduced what came to be called **algebra** (from *al-jabr*). Another book on arithmetic was later translated into Latin and, from this one, his own name produced the word **algorithm**. Yet another book described a new number system and was the source used by Western mathematicians several centuries later. Because of its origins (with the Hindus of India) and its transmission through the Arabs it is known as the **Hindu–Arabic number system**.

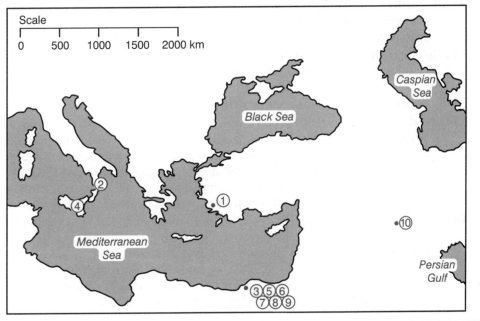

mathematicians of later times

Adelard of Bath (c.1075–1160) was an English scholar. At this time the main language of mathematical texts was Arabic. With the gradual emergence of the Western world from the Dark Ages there was a great need for these texts to be translated. Adelard was one of the first to do so, translating several Arabic texts into Latin which was the principal language of scholars for several centuries to come. Unfortunately his work did not become as well known as it deserved to be.

Fibonacci (1170–1250) was an Italian mathematician who travelled extensively and who published (in 1202) the *Liber Abaci* which made the work of **al Khwarizmi** more widely known to the Western world. In that book, through one of his problems, he introduced the **Fibonacci sequence**.

Nicolo Tartaglia (1500–1557) was an Italian mathematician who is generally credited with being the first to solve all types of cubic equations using only algebraic methods. Previous to that only a few particular types could be solved. He confided his method to **Cardano** under an oath of secrecy.

Girolamo Cardano (1501–1576) was an Italian mathematician who had considerable skill as a doctor of medicine. He was also a rogue and a gambler. His book *Ars Magna* was a major work on algebra in which he gave **Tartaglia's** method for solving cubics – breaking his oath of secrecy. In that book he gave a hint of the existence of **imaginary numbers**. He later published the first book on probability.

Robert Recorde (1510–1558) was an English mathematician and was the first writer in English of books on arithmetic, geometry and algebra. He is regarded as the founder of mathematics in England. His book *The Whetstone of Witte* introduced the equals sign (=) using two parallel lines of the same length as he said, 'because no 2 things can be more equal'.

John Napier (1550–1617) was a Scottish laird with a keen interest in mathematics and particularly in methods of computation. He invented a primitive calculating machine, **Napier's rods** and **logarithms**.

Marin Mersenne (1588–1648) was a French friar who was very interested in mathematics and particularly in **number theory**. He was in touch with many mathematicians and scientists of that time, such as **Descartes** and **Fermat**. He devised **Mersenne primes**.

Pierre de Fermat (1601–1665) was a lawyer working for the French government. His hobby was mathematics and he was probably the greatest amateur mathematician of all time. His especial interest was **number theory**. He is remembered particularly for what is known as **Fermat's last theorem**. The proof of this eluded mathematicians for over 300 years but, in their attempts to find such a proof, much other useful mathematics was generated.

Blaise Pascal (1623–1662) was a French mathematician, physicist and theologian. He was a child prodigy who worked out the principles of **Euclidean geometry** by himself and invented, and built, a mechanical calculating machine. He is credited, together with **Fermat**, with being the founder of probability theory. He is remembered in **Pascal's triangle**, which he published, though it had been known about for over 500 years by then.

Sir Isaac Newton (1643–1727) was an English mathematician and scientist who is generally thought to be one of the greatest mathematicians of all time. He identified the principle of gravitation and the fact that it applied to all bodies throughout the Universe, establishing a formula to predict its effect in all circumstances. He formulated the three laws of motion and, by using a prism, established that white light was made up of a spectrum of colours. One of his greatest achievements was the invention of the **calculus**.

Gottfried von Leibniz (1646–1716) was a German mathematician who, independently of **Newton**, but about the same time, also invented the **calculus**. Though their methods were the same in principle, they differed widely in the notation they used. Controversy over which was the better dragged on for almost a century, but it is the Leibniz notation we use today.

Leonhard Euler (1707–1783) was a Swiss mathematician who spent the major part of his working life in Russia. He is generally considered to have been the most prolific of all mathematicians, and it was not until many years after his death that all his writings had appeared in print. He had an incredible memory which he demonstrated to good effect in the last 17 years of his life when he was totally blind, but continued to generate as much mathematics as when he could see. He invented the topic we now know as **topology**.

Carl Frederick Gauss (1777–1855) was a German mathematician who, along with **Archimedes** and **Newton**, is considered one of the all-time 'greats'. He was a child prodigy who, at the age of three, showed great aptitude for work with numbers and this stayed with him throughout his life. He did original work in many branches of mathematics, but particularly in algebra and arithmetic. He was the first to prove the **fundamental theorem of arithmetic**.

Henri Poincaré (1854–1912), a French mathematician, was the last mathematician able to work equally well on all aspects of mathematics, and produced a wide variety of important results. He is often described as the 'last great universalist'. Since those times, mathematicians have tended to specialise in particular branches of mathematics.

David Hilbert (1862–1943) was a German mathematician who established a new approach to geometry with his book *Foundations of Geometry* (1899). Then, in a famous talk in Paris in 1900 he posed 23 problems, most of which he saw as being important to the future development of mathematics. Many of those problems are still of interest to mathematicians.

Bertrand Russell (1872–1970) achieved fame as a philosopher but, up to the age of 38, his main interest was mathematics. In response to **Hilbert's** second problem he attempted in the book *Principles of Mathematics* (1910) to establish the rules of arithmetic in a rigid logical system.

Kurt Gödel (1906–1978) was an Austrian mathematician who disturbed mathematicians worldwide with his surprising proof, in 1931, that in any system based only on logic, it is always possible to formulate some theorems that cannot be proved. Sadly, such theorems cannot always be identified and, unsurprisingly, this continues to trouble mathematicians.

matrices

array An array is an orderly display of data arranged in a rectangular shape.

element An element of an **array** is one complete piece of the data in the array.
> *Example: In the array* $\begin{array}{cc} 3.8 & 4.2 \\ 6.1 & 7.9 \end{array}$ *3.8 is an element, but 8 is not as it is not the complete piece of data.*

matrix A matrix is a rectangular **array** of **elements**. *Usually the elements are all of the same type (numbers, symbols, algebraic expressions etc.) and the array is enclosed in either square or round brackets. The individual elements are separated only by spaces; commas or other dividing marks are not used.*
> *Example:* $\begin{pmatrix} 1 & 4 & ^-3 \\ 6 & 1.5 & 7 \end{pmatrix}$ $\begin{pmatrix} A & X \\ Y & B \end{pmatrix}$ $\begin{pmatrix} 3x+8 & y+1 \\ 3y-4 & 2x-7 \end{pmatrix}$

row A row of a **matrix** is the set of **elements** making up one complete line reading across the matrix from left to right.
> *Example: The matrix* $\begin{pmatrix} 1 & 7 \\ 8 & 5 \end{pmatrix}$ *has 2 rows: (1 7) and (8 5)*

column A column of a **matrix** is the set of **elements** making up one complete line reading down the matrix from top to bottom.
> *Example: The matrix* $\begin{pmatrix} 4 & 9 & 6 \\ 1 & 0 & 7 \end{pmatrix}$ *has 3 columns.*

order The order of a **matrix** is a measure of its size, stating it as 'the number of **rows** by the number of **columns**'. *They must be stated that way round.*
> *Example:* $\begin{pmatrix} 7 & ^-6 & 0 & 3 \\ 9 & 1 & 8 & 3 \end{pmatrix}$ *is a 2 by 4 matrix.*

square matrix A square matrix has the same number of **rows** and **columns**.
> *Example:* $\begin{pmatrix} 4 & ^-1 \\ 0 & 6 \end{pmatrix}$ *is a square matrix. It is also a 2 by 2 matrix.*

row matrix A row matrix has only a single **row**.
> *Example:* $(3 \quad ^-4 \quad 10)$ *is a row matrix. It is also a 1 by 3 matrix.*

column matrix A column matrix has only a single **column**.

diagonals The set of **elements** making up one complete line reading from the top left corner to the bottom right corner of a **square matrix** is the **leading, main** or **principal diagonal**. *The other line (from top right to bottom left) is the* **secondary** *or* **trailing diagonal**.
> *Example: In* $\begin{pmatrix} 7 & 2 \\ 4 & ^-9 \end{pmatrix}$ *the main diagonal is* $7 \quad ^-9$ *and the trailing diagonal is* $2 \quad 4$

trace The trace of a **square matrix** is the sum of the elements in the **main diagonal**
> *Example: In* $\begin{pmatrix} ^-6 & 10 \\ 14 & 18 \end{pmatrix}$ *the trace is 12 (= $^-6$ + 18)*

transpose The transpose of a **matrix** is made by rewriting the **rows** of the matrix as **columns**. *In a square matrix the leading diagonal will be unchanged.*
> *Example: For* $\begin{pmatrix} a & b & c \\ d & e & f \end{pmatrix}$ *the transpose is* $\begin{pmatrix} a & d \\ b & e \\ c & f \end{pmatrix}$

addition Two matrices may be added to make a new matrix, provided they are of the same **order**, by adding corresponding **elements** in each to form elements for the new matrix. *The new matrix will be of the same order.*

Example: $\begin{pmatrix} 1 & 2 \\ 3 & 4 \end{pmatrix} + \begin{pmatrix} 5 & 9 \\ 7 & 8 \end{pmatrix} \rightarrow \begin{pmatrix} 1+5 & 2+9 \\ 3+7 & 4+8 \end{pmatrix} = \begin{pmatrix} 6 & 11 \\ 10 & 12 \end{pmatrix}$

scalar multiplication A scalar is a number which, when written in front (to the left) of a **matrix**, means that all the **elements** of that matrix have to be multiplied by that number.

Example: $3 \begin{pmatrix} 2 & 0 \\ 1 & 5 \end{pmatrix} \rightarrow \begin{pmatrix} 3 \times 2 & 3 \times 0 \\ 3 \times 1 & 3 \times 5 \end{pmatrix} = \begin{pmatrix} 6 & 0 \\ 3 & 15 \end{pmatrix}$

multiplication Two matrices may be multiplied, provided that the number of COLUMNS in the FIRST matrix is the same as the number of ROWS in the SECOND matrix. *It is done by laying each row of the first matrix against each column of the second matrix, multiplying the pairs of elements, and adding the results together to make a single element for the answer matrix. Matrix multiplication depends on the ORDER in which the two are written. Changing the order may give a different answer, or multiplication may not be possible.*
Example:

$\begin{pmatrix} 4 & 2 \\ 3 & 1 \end{pmatrix}\begin{pmatrix} 5 & 7 & 8 \\ 6 & 0 & 9 \end{pmatrix} \rightarrow \begin{pmatrix} 4 \times 5 + 2 \times 6 & 4 \times 7 + 2 \times 0 & 4 \times 8 + 2 \times 9 \\ 3 \times 5 + 1 \times 6 & 3 \times 7 + 1 \times 0 & 3 \times 8 + 1 \times 9 \end{pmatrix} = \begin{pmatrix} 32 & 28 & 50 \\ 21 & 21 & 33 \end{pmatrix}$

But $\begin{pmatrix} 5 & 7 & 8 \\ 6 & 0 & 9 \end{pmatrix}\begin{pmatrix} 4 & 2 \\ 3 & 1 \end{pmatrix}$ *cannot be done.*

diagonal matrix A diagonal matrix is a **square matrix** which has all its **elements** equal to zero, except for those on the **main diagonal**.

identity matrix The identity matrix (for multiplication) is a **square matrix** whose **elements** in the **main diagonal** are all 1's, and the others are all zero.

Example: $\begin{pmatrix} 1 & 0 \\ 0 & 1 \end{pmatrix}$ *is the 2 by 2 identity matrix (for multiplication).*

determinant The determinant of a **square matrix** is a single number obtained by applying a particular set of rules to the **elements** of that matrix.

$\begin{vmatrix} a & b \\ c & d \end{vmatrix}$ The two ruled lines are a symbol meaning that the **determinant** of the matrix $\begin{pmatrix} a & b \\ c & d \end{pmatrix}$ has to be found.

For a 2 by 2 matrix the rule is $ad - bc$

Example: $\begin{vmatrix} 2 & 3 \\ 4 & 7 \end{vmatrix}$ is $(2 \times 7) - (3 \times 4) = 14 - 12 = 2$

singular matrix A singular matrix is a **square matrix** with **determinant** = zero.

inverse matrix The inverse, for multiplication, of a **square matrix** (which must NOT be **singular**) is another matrix such that when the two are multiplied together, in any order, then the result is the **identity matrix**. *Example:*

The inverse of $\begin{pmatrix} 2 & 1 \\ 5 & 3 \end{pmatrix}$ is $\begin{pmatrix} 3 & -1 \\ -5 & 2 \end{pmatrix}$ since $\begin{pmatrix} 2 & 1 \\ 5 & 3 \end{pmatrix}\begin{pmatrix} 3 & -1 \\ -5 & 2 \end{pmatrix} = \begin{pmatrix} 1 & 0 \\ 0 & 1 \end{pmatrix}$

direction For movement over the surface of the Earth the direction (of a line) is measured relative to another line pointing to a position called North. *There are two positions known as North. The North Pole is fixed and known as True North. A magnetic compass points at Magnetic North. The difference between True North and Magnetic North varies and can be as much as 20°.*

points of the compass The points of the compass are those directions defined (relative to North) by dividing a circle into 4, 8, 16 or 32 equal parts. *The principal points are North, South, East and West, and the next four are the points between those: North East (NE), etc.*

compass angles are the angles (in degrees) measured CLOCKWISE from the North-line to the line of the required direction. *This allows a direction to be given to any (possible) accuracy between 0° and 360°. Such angles are always written with 3 digits, so 57° becomes 057° and 6° is 006°.*

bearing The bearing of position B FROM position A is the **direction** (usually given as a **compass angle**) in which someone travelling in a straight line FROM A TO B must go. *Care should be taken when using 'from' and 'to' in this work.*

reciprocal bearing A reciprocal bearing is the direction which is the reverse of the given **bearing**. *If the given bearing is FROM A TO B, the reciprocal bearing is the bearing FROM B TO A.*

> If bearing is LESS than 180°, ADD 180° to get the reciprocal.
> If bearing is MORE than 180°, SUBTRACT 180° to get the reciprocal.

back bearing = **reciprocal bearing**

pole A pole is one of two positions (one North, one South) on the Earth's surface through which the **axis of rotation** of the Earth passes.

great circle A great circle is any circle drawn on the surface of a sphere (the Earth) whose centre is at the centre of the sphere. *All great circles on a sphere are of the same size, and any one divides the sphere into two hemispheres.*

small circle A small circle is any circle drawn on the surface of a sphere (the Earth) that is NOT a **great circle**.

equator The equator is the **great circle** around the Earth that is perpendicular to the **axis of rotation**. *It is equidistant from either pole, and divides the Earth into northern and southern hemispheres.*

latitude A line of latitude is a **small circle** on the Earth's surface, parallel to the **equator**, whose position is given North or South of the equator.

meridian A meridian is half of a **great circle** on the Earth's surface going from one **pole** to the other. *It appears as a line running North and South, crossing the equator at right angles.*

Greenwich Meridian The Greenwich meridian is the **meridian** which passes through a fixed point in the old Greenwich Observatory (in London).

longitude A line of longitude is a **meridian** whose position is given, East or West, relative to the **Greenwich Meridian** as measured by the angle at the centre of the Earth between the **great circles** forming those two meridians.

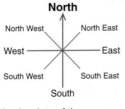
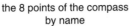

the 8 points of the compass
by name

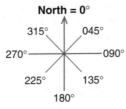

the 8 points of the compass
measured in degrees
clockwise from North

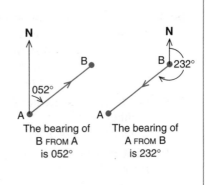

The bearing of
B FROM A
is 052°

The bearing of
A FROM B
is 232°

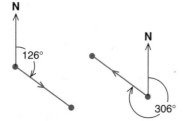

A bearing of 126° (which is < 180) has
a reciprocal bearing of 126 + 180 = 306°

OR

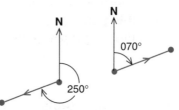

A bearing of 250° (which is > 180) has
a reciprocal bearing of 250 − 180 = 070°

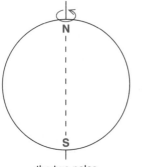

the two poles
and the axis of rotation

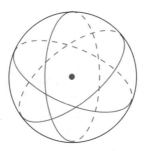

great circles on a sphere
and their common centre

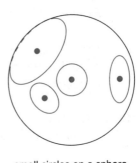

small circles on a sphere
and their different centres

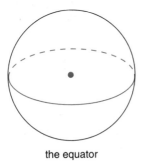

the equator

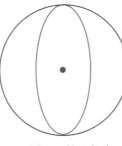

meridians of longitude

circles of latitude

perfect numbers are numbers whose **proper factors** add up to the number itself.
> *Examples: 6 is a perfect number since $1 + 2 + 3 = 6$*
> *28 is a perfect number since $1 + 2 + 4 + 7 + 14 = 28$*
> *The next three are 496, 8128 and 33 550 336*
> *No odd perfect numbers are known.*

deficient numbers are numbers whose **proper factors** add up to LESS THAN the number itself.
> *Example: 16 is a deficient number since $1 + 2 + 4 + 8 = 15$*

abundant numbers are numbers whose **proper factors** add up to MORE THAN the number itself.
> *Example: 20 is an abundant number since $1 + 2 + 4 + 5 + 10 = 22$*

amicable pair An amicable pair of numbers is two numbers with the property that the **proper factors** of each one add up to the value of the other.
> *Example: 220 and 284 are an amicable pair.*
> *220 gives $1 + 2 + 4 + 5 + 10 + 11 + 20 + 22 + 44 + 55 = 284$*
> *284 gives $1 + 2 + 4 + 71 + 142 = 220$*

automorphic numbers are numbers whose last digits are unchanged after the number has been squared.
> *Example: 76 and 625 are automorphic since $76^2 = 5776$; $625^2 = 390625$*

palindrome A palindrome is a number (or word) which is unchanged whether it is read from left to right or from right to left.
> *Examples: 77, 565, 34843, 1962691 are all palindromic numbers.*
> *11, 727, 36563, 9714179 are all palindromic primes.*

pandigital A pandigital number or expression is one which contains each of the digits 1 to 9 (or 0 to 9) once and once only.
> *Examples: $26 + 48 + 79 = 153$ is a pandigital expression.*
> *139 854 276 and 9 814 072 356 are pandigital squares.*

Harshad numbers are numbers which can be divided exactly by their **digit sum**.
> *Example: 1729 is a Harshad number since its digit-sum $(1 + 7 + 2 + 9)$*
> *is 19, and 19 divides exactly into 1729*

Kaprekar's constant $= 6174$. This is the result eventually produced by carrying out the following operations. Make a 4-digit number by using at least 2 different, non-zero, digits. Place the digits in order: largest to smallest and smallest to largest, to make two other numbers. Subtract the smaller from the larger to make a new number. Continue repeating this process until 6174 is obtained.
> *Example: Starting with 1998:* $1998 \rightarrow 9981 - 1899 = 8082$
> $8082 \rightarrow 8820 - 0288 = 8532$
> $8532 \rightarrow 8532 - 2358 = \mathbf{6174}$

cycles using numbers are found by first making a rule by which one number is used to produce another number. Then, by continuous use of that rule, seeing if a loop or chain is made when a previous number is repeated. *By using the rule given to make **happy numbers** some cycles will be found.*

partition To partition a number is to break it up into a separate set of numbers which add up to make the original number. *Whole numbers are used throughout. Zero is not used but the number itself is included. Merely reordering the set is not considered to represent a different partition. The number of different ways in which each of the numbers 1 to 10 may be partitioned are shown in the table on the right.* *Example: 5 can be partitioned in 7 different ways as:*

Number	Ways
1	1
2	2
3	3
4	5
5	7
6	11
7	15
8	22
9	30
10	42

$1 + 1 + 1 + 1 + 1$ $1 + 1 + 1 + 2$ $1 + 1 + 3$
$1 + 2 + 2$ $1 + 4$ $2 + 3$ 5

persistence The digits of a number are multiplied together to make another number. This process is continued on each new number until only a single digit is obtained. The number of times the process has to be repeated to achieve this is a measure of the persistence of the original number.
Example: $79 \rightarrow 63 \rightarrow 18 \rightarrow 8$ so 79 has a persistence of 3

polite numbers have been defined as those numbers which can be made by adding together two or more consecutive whole numbers. *This can often be done in more than one way, and the number of ways it can be done is a measure of the politeness of a number.*
Example: $15 = 1 + 2 + 3 + 4 + 5$ and $4 + 5 + 6$ and $7 + 8$, so 15 is
a polite number and has a politeness of 3

happy numbers A number has all its digits squared and added together to make a new number. This process is repeated until a 1 is obtained, when the original number is described as happy. *If a 1 is never obtained then the original number is said to be 'sad'.*
Example: $19 \rightarrow 1^2 + 9^2 = 82 \rightarrow 8^2 + 2^2 = 68 \rightarrow 6^2 + 8^2 = 100 \rightarrow 1^2 = 1$
so 19 is a happy number.

cutting numbers have been defined as that sequence of numbers produced when a given shape is cut up into the maximum number of pieces by a succession of 1, 2, 3, 4 . . . . cuts and without re-arranging the pieces between cuts.
Examples: A line makes 2, 3, 4, 5 . . . pieces after 1, 2, 3, 4 . . . cuts.
A circle produces 2, 4, 7, 11 . . . pieces after 1, 2, 3, 4 . . . cuts.

repunits are numbers made up only of 1's (= repeated units). *A short way of writing such numbers is I_n where n is the number of 1's to be used.*
Examples: $I_3 = 111$ $I_6 = 111111$ I_2 I_{19} I_{23} are all primes.

multigrades A multigrade is an equality between two expressions, each requiring some numbers to be raised to a power and added, which is true for more than one value of that power.
Examples: $1^n + 2^n + 6^n = 4^n + 5^n$ $(n = 1$ or $2)$
$1^n + 6^n + 8^n = 2^n + 4^n + 9^n$ $(n = 1$ or $2)$

digital invariants have been defined as those numbers which are equal to the sum of all their separate digits when raised to the same power.
Example: 153 is a digital invariant since $1^3 + 5^3 + 3^3 = 1 + 125 + 27 = 153$

superscript A superscript is a letter or number written in small type placed to the right and at the top of a letter, number or symbol written full size.
Example: In A^2 A^n 5^2 7^x y^3 ^{2} ^{n} ^{2} ^{x} and ^{3} are all superscripts.

subscript A subscript is a letter or number written in small type placed to the right and at the bottom of a letter, number or symbol written in full size.
Example: In A_1 A_3 B_n x_2 x_r $_1$ $_3$ $_n$ $_2$ and $_r$ are all subscripts.

index notation is a way of indicating how a number (or symbol) must be operated on by using another number written as a **superscript** to the first; this second number is called an index. *When the index is a positive whole number that number indicates how many of the first number or symbol must be multiplied together. When the index is a fraction then it indicates a root has to be found.*
Examples: $A^2 = A \times A$ $5^3 = 5 \times 5 \times 5 = 125$ $9^{\frac{1}{2}} = \sqrt{9} = 3$

base In **index notation** the base is the number (or symbol) upon which the **index** is to operate.
Examples: In A^2 y^{-1} 3^4 10^x A, y, 3 and 10 are all bases

positive index A positive index is an **index** which is greater than zero.

zero index A zero index is an **index** which is equal to zero. *ANY number raised to a zero index is equal to 1*
Examples: $1^0 = 1$; $2^0 = 1$; $99^0 = 1$; $x^0 = 1$ BUT $0^0 = 0$

negative index An **index** having a negative value indicates a **reciprocal** has to be taken AFTER the index has been applied to the **base**.
Examples: $x^{-1} = \dfrac{1}{x}$ $x^{-2} = \dfrac{1}{x^2}$ $2^{-3} = \dfrac{1}{2^3} = \dfrac{1}{8}$

power $\equiv$ **index**

exponent $\equiv$ **index**

standard form is a way of displaying a number in the form of a first number, whose value lies between 1 and 10, and a second number which is always 10 with a suitable **index**, so that the two numbers multiplied together equal the value of the intended number.
Examples: 1.436×10^2 means 143.6 $1.436 \times 10^{-3} = 0.001436$

scientific notation $\equiv$ **standard form**. *This is the description usually found on electronic calculators where 1.436×10^2 would appear as 1.436E02*

reciprocal The reciprocal of a number is the value given by dividing 1 BY that number, or dividing that number INTO 1
Examples: Reciprocal of 2 is $\dfrac{1}{2}$ or $1 \div 2 = 0.5$

of 7 it is $\dfrac{1}{7} = 0.142857 \dots$

factorial The value of factorial n is found by multiplying together all the whole numbers from 1 up to, and including, n. A special case is $0! = 1$
Example: Factorial 5 is $5 \times 4 \times 3 \times 2 \times 1 = 120$

! when written AFTER the number n is the symbol for **factorial** n. *Example: $5! = 120$*

0!	=	1
1!	=	1
2!	=	2
3!	=	6
4!	=	24
5!	=	120
6!	=	720
7!	=	5 040
8!	=	40 320
9!	=	362 880
10!	=	3 628 800

permutation A permutation of a set of objects is an ORDERED arrangement of those objects. *The fact it is ordered means that* AB *is considered to be different from* BA. *Interest is usually focused on how many different permutations are possible from a given set of objects.*
Example: Set ABC *has 6 permutations:* ABC, ACB, BAC, BCA, CAB, CBA

With n different objects there are $n!$ permutations possible.

$^n P_r$ is the symbol for the total number of **permutations** possible when, from a set of n objects, r are chosen at a time.

When the objects are all distinguishably different then $^n P_r = \dfrac{n!}{(n-r)!}$

combination A combination of objects is an UNORDERED arrangement of those objects. *The fact it is unordered means that* ABC *is considered to be the same as* BCA *or* CAB *or* CBA, *etc.*

$^n C_r$ is the symbol for the total number of **combinations** possible when, from a set of n objects, r are chosen at a time. *Said as 'From n choose r'*

When the objects are all distinguishably different then $^n C_r = \dfrac{n!}{(n-r)!\, r!}$

When $r = n$ then $^n C_r = 1$

Example: $^{49}C_6 = \dfrac{49!}{43! \times 6!} = 13\,983\,816$ *(which is almost 14 million)*

Some values of $^n C_r$ for various values of n and r											
$n =$	2	3	4	5	6	7	8	9	10	11	12
$r = 2$	1	3	6	10	15	21	28	36	45	55	66
3		1	4	10	20	35	56	84	120	165	220
4			1	5	15	35	70	126	210	330	495
5				1	6	21	56	126	252	462	792
6					1	7	28	84	210	462	924
7						1	8	36	120	330	792
8							1	9	45	165	495

dozen $\equiv 12$

gross $\equiv 144$ $(= 12 \times 12$ or a dozen dozens)

thousand $\equiv 1000$ or 10^3

million $\equiv 1000\,000$ or 10^6 $\equiv$ a thousand thousands

billion $\equiv 1000\,000\,000$ or 10^9 $\equiv$ a thousand millions
In the older British system a billion was taken to be a million millions but this has died out and the International system, given here, is now followed.

trillion $\equiv 1000\,000\,000\,000$ or 10^{12} $\equiv$ a thousand billions

Other numbers in this series can be named by changing the prefix to indicate how many thousands of thousands there are.
Example: quad $\equiv 4$ so a quadrillion has 1000^4 thousands $\equiv 10^{15}$

centillion $\equiv 1000^{100} \times 1000$ or 10^{303}

googol A googol is 10^{100} which is 1 followed by 100 zeros.

googolplex A googolplex is 10^{googol} which is 1 followed by a googol of zeros.

number system A number system is made up of a set of defined symbols and the numbers they represent, together with rules for forming larger numbers from those symbols.

Hindu–Arabic number system The Hindu–Arabic number system is the **number system** which is used for ordinary **arithmetic**; and the symbols are the **digits** 0 to 9

place-value A place-value **number system** is one in which the positions of the symbols affect the overall value of the number. *Early number systems did not use a place value system and it is sometimes difficult to decide what the value of the number is meant to be.*
Example: In the number 7361 the symbol 3 has a value of 3-hundreds, but in the number 4138 it has a value of 3-tens.

base The base of a **place-value number system** controls the relationship between the places. *Usually it is also the number of different symbols used. Example: In ordinary arithmetic the base is 10, and there are 10 symbols used to make numbers. They are 0, 1, 2, 3, 4, 5, 6, 7, 8 and 9.*

binary A binary **number system** uses a **base** of 2. *The 2 symbols used are 0 and 1. Numbers in this base look like 1001101 ($\equiv$ 77 in decimal).*

ternary A ternary **number system** uses a **base** of 3. *The 3 symbols used are 0, 1 and 2. Numbers in this base look like 1020211 ($\equiv$ 913 in decimal).*

octal An octal **number system** uses a **base** of 8. *The 8 symbols used are 0, 1, 2, 3, 4, 5, 6 and 7. This system is used in some computer work.*

decimal A decimal **number system** uses a **base** of 10. *This is the system used in ordinary arithmetic.*

denary $\equiv$ **decimal**.

hexadecimal A hexadecimal **number system** uses a **base** of 16. *The 16 symbols used are 0, 1, 2, 3, 4, 5, 6, 7, 8, 9, A, B, C, D, E and F. This system is used widely in computer work. To signal that a hexadecimal number is intended it is usually prefaced with the symbol &. A signal is necessary because, for instance, the hexadecimal value of 10 is decimal 16.*
Example: The largest two-symbol number in the hexadecimal system is FF (written &FF) which is equivalent to 255 in the decimal system.

place-value headings The column headings needed for any **place-value system** are fixed by the **base** used in that system. The way in which this is done is shown, where b is the value of the base. *Decimal values for the most common bases are given. The table can be extended to the right to form fractions in any system, especially decimal. The values are then b^{-1}, b^{-2}, b^{-3}, etc. giving rise to one-tenth, one-hundredth, etc. for decimals.*

		b^3	b^2	b	units
binary system	$b = 2$	8	4	2	1
ternary system	$b = 3$	81	9	3	1
octal system	$b = 8$	512	64	8	1
decimal system	$b = 10$	1000	100	10	1
duodecimal	$b = 12$	1728	144	12	1
hexadecimal	$b = 16$	4096	256	16	1

place-value names are those used in the decimal system to name the columns as shown below. *Only in the decimal system are the columns named. Note how hundreds, tens and units (h t u) are repeated under each major name.*

trillions			billions			millions			thousands			hundreds	tens	units
h	t	u	h	t	u	h	t	u	h	t	u			

place-holder or zero The place-holder in the **Hindu–Arabic number system** is the symbol for zero ($\equiv 0$). *It is necessary to have such a symbol in a place-value system or else it would be impossible to know in which column each of the other symbols should be placed, and so know its true value. Example: In 3024 the 3-symbol is valued at 3-thousands. If the 0 was left out as there were no hundreds, then 342 would value the 3-symbol at 3-hundreds.*

additive number systems are **number systems** in which the bigger numbers are formed by using enough of the basic symbols to add up to the number required. *With such systems the symbols can be placed in any order; it is only necessary to make clear to which group, or number, each symbol belongs.*

Egyptian number system In ancient Egypt they used an **additive number system**, using the symbols

$|\equiv 1$ $\cap \equiv 10$ $\mathcal{9} \equiv 100$ $\overset{\mathcal{D}}{\underset{\Delta}{\int}} \equiv 1000$ $\mathcal{(}\equiv 10000$

Example:

$\begin{matrix} || \cap\cap \\ ||| \cap \end{matrix} \mathcal{99}\overset{\mathcal{D}}{\underset{\Delta}{\int}} \equiv 5 + 30 + 200 + 1000 = 1235$

early Roman number system The Romans mainly used an **additive number system**, using the symbols

$I \equiv 1$ $V \equiv 5$ $X \equiv 10$ $L \equiv 50$ $C \equiv 100$ $D \equiv 500$ $M \equiv 1000$

Examples: IIII $\equiv 4$ XXXX $\equiv 40$ LXXXX $\equiv 90$ CCCXX $\equiv 320$

later Roman number system Sometimes, to save space, the Romans used a subtractive idea in their number system: symbols now HAD to be written in size order from left to right but, if a smaller one preceded a larger one then the smaller had to be subtracted from the larger; this could only be applied to an adjacent pair of symbols. *This idea was little used by the Romans and was only applied to every number in comparatively recent times. Examples: IV (is 1,5 so 5 – 1) $\equiv 4$ VC (is 5,100 so 100 – 5) $\equiv 95$ IX $\equiv 9$ XLIV $\equiv 44$ MCMXLIV $\equiv 1944$ MCMXCV $\equiv 1995$*

Greek number system Ancient Greeks used an **additive number system**, using their alphabet as number symbols.

Example: σ π θ (is 200 + 80 + 9) $\equiv$ 289

Babylonian number system The Babylonians used a mixture of an **additive number system**, with some subtraction, and a **place-value system** with a **base** of 60. *This base lingers on in our measurement of time and angle (60 minutes in 1 hour etc.). The Babylonians needed only three symbols: for 1, 10, and subtraction. They did not have a place-holder.*

natural numbers are the set of numbers 1, 2, 3, 4, 5, 6, . . . as used in counting. *It is a matter of choice whether 0 is included or not.*

counting numbers ≡ **natural numbers**.

integers are numbers made from the **natural numbers** (including 0) by putting a positive or a negative sign in front. *The positive sign is often omitted. The integers are:* ..., ⁻5, ⁻4, ⁻3, ⁻2, ⁻1, 0, 1, 2, 3, 4, 5, ...

whole numbers is a term used rather loosely to mean EITHER the **natural numbers** OR the **integers**; it depends upon the context.

positive integers ≡ **natural numbers** (excluding zero).

signed numbers ≡ **integers**.

directed numbers ≡ **integers**.

rational numbers can be written in the form $\frac{a}{b}$ where a and b are both **integers** and b is not zero.

Examples: ⁻4.5 $1\frac{1}{3}$ 0.0909090909... $3.\dot{1}4285\dot{7}$ 8

are all rational since they can be re-written as:

$$\frac{^-9}{2} \qquad \frac{4}{3} \qquad \frac{1}{11} \qquad \frac{22}{7} \qquad \frac{8}{1}$$

irrational numbers can only be written in number form (using no symbols) as a never-ending, non-repeating decimal fraction. *An irrational number CANNOT be written in the form of a rational number.*

Example: 1.234567891011121314151617181920212223 24...

To generate this number it is only necessary to write out the natural numbers and it could go on for ever without repeating. It is an irrational number. The square root of any prime number is irrational, as also is π

real numbers The set of real numbers is made up of all the **rational** and **irrational** numbers together.

positive numbers are all the **real numbers** which are greater than 0.

negative numbers are all the **real numbers** which are less than 0.

$\mathbb{N}$ is the symbol to indicate that the set of **natural numbers** is to be used.

$\mathbb{Z}$ is the symbol to indicate that the set of **integers** is to be used.

$\mathbb{Q}$ is the symbol to indicate that the set of **rational numbers** is to be used.

$\mathbb{R}$ is the symbol to indicate that the set of **real numbers** is to be used.

$\mathbb{C}$ is the symbol to indicate that the set of **complex numbers** is to be used.

number line A number line is a graduated straight line along which it is possible *(in theory)* to mark ALL the **real numbers**.

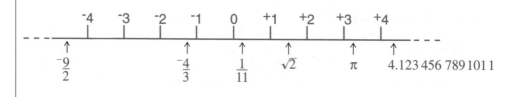

cardinal numbers are the **natural numbers** used to describe 'how many' objects there are in a set.
Example: 'The set R B Z N F B R X *has 8 letters altogether but only 6 are different.' In that statement 8 and 6 are used as cardinal numbers.*

ordinal numbers are the **natural numbers** used to describe the 'position' of an object in a set which arranged in order. *Most common are 1st, 2nd, 3rd etc. Example:'The set* C N B I Z K F R G *has* R *as the 8th letter and in position 6 there is a* K*.' In that statement 8 and 6 are used as ordinal numbers.*

identification numbers are numbers which are neither **cardinal** nor **ordinal** but which are given to objects or persons to help distinguish them in some way. *They may be part-cardinal and/or part-ordinal when they are made, but that usually has nothing to do with their use.*
Examples: a number 9 bus; National Insurance numbers; Personal Identification Numbers (PIN); catalogue numbers etc.

i is the symbol for $\sqrt{-1}$ (the square root of negative one) which cannot exist as a **real number**. **j** is also used, especially in science and engineering.

imaginary numbers are the square roots of negative numbers. *They are called imaginary since the square root of a negative number cannot be* **real**. *The way in which they are expressed is based on this argument:*
$\sqrt{-k}$ *can always be re-written as* $\sqrt{k} \times \sqrt{-1}$ *and k is positive*
so $\sqrt{k}$ *is real.*
Suppose $\sqrt{k} = b$ *then* $\sqrt{-k} = \sqrt{k} \times \sqrt{-1} = b\mathbf{i}$
So any imaginary number can be written in the form **bi***, where* **b** *is real.*

complex numbers involve a combination of **real** and **imaginary numbers**. *They are written in the form* **a + bi***, where* **a** *and* **b** *are real numbers.*

Argand diagram An Argand diagram allows **complex numbers** to be shown in in a way that is not possible on a simple real **number line**. *The diagram uses a number line as one axis to plot the real part of the complex number, and a second number line to form an axis at right angles to the first, on which the imaginary part of the same complex number can be plotted. With these two points as a coordinate pair, a single point can be plotted to represent the complex number. Operations on complex numbers can also be shown.*
Example showing the complex number 4 + 3i on an Argand diagram:

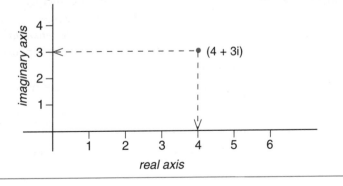

patterns

pattern A pattern may be EITHER something that is to serve as a model to be copied OR, more usually in mathematics, a set of objects or elements arranged in order according to a rule. *Repetition, in some way, underlies all patterns.*

motif In a spatial **pattern** the motif is the basic object or element which is repeated to make the pattern. *A motif can be a single object or made up of a collection of objects. The actual drawing used for the motif is of no importance at all – it can be a triangle, a flower, a face, a complete picture even – so long as it really is the basic element which is repeated throughout the pattern. Examples:*

mapping a pattern A spatial **pattern** is said to be mapped on to itself if a **translation, reflection, rotation, glide reflection**, or any combination of them, acting on the whole pattern, serves to place each **motif** of the pattern exactly onto another copy of that motif – so the pattern matches with itself. *For the purpose of mapping a pattern it is important to assume that the pattern goes on indefinitely.*

frieze patterns A frieze pattern is a spatial **pattern** which 'grows' in one dimension only, and the pattern can be **mapped** onto itself. *The only possible rotation for a frieze pattern is a half-turn. There are just **7** different frieze patterns and all of them are shown on the opposite page. They are also known as **border** patterns and **strip patterns**.*

wallpaper patterns A wallpaper pattern is a spatial **pattern** which 'grows' in two dimensions and the pattern can be **mapped** on to itself. *The only possible rotations for a wallpaper pattern are through 180°, 120°, 90° and 60°. There are just **17** different wallpaper patterns. Care needs to be taken in analysing these patterns that the full extent of the motif is properly identified. Any **periodic tiling** is an example of a wallpaper pattern.*

three-dimensional patterns Just as patterns can be made having one dimension (frieze patterns) and two dimensions (wallpaper patterns), other patterns can be made having three dimensions which can be mapped onto themselves. *There are just **230** different three-dimensional patterns. They have all been identified and classified, and play an important part in the scientific study of crystals.*

number patterns are usually different from spatial patterns in that they do not have a motif in the same way that a spatial pattern does. The repetition that allows the word 'pattern' to be used is to be found in the rule (or rules) which control the making of the numbers. *Examples of number patterns are to be found in many **sequences** and arrangements such as **Pascal's triangle**.*

Frieze Patterns

All 7 possible (different) ways of making a frieze pattern are shown below.
The transformations given at the right of each pattern are those which define the type.
In addition to the transformations given, all patterns can be mapped on to themselves by means of a translation. All these patterns start with the same motif.

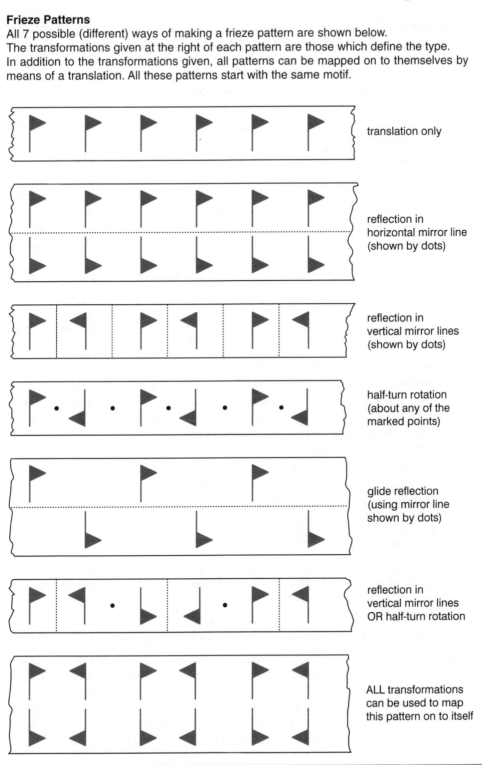

translation only

reflection in horizontal mirror line (shown by dots)

reflection in vertical mirror lines (shown by dots)

half-turn rotation (about any of the marked points)

glide reflection (using mirror line shown by dots)

reflection in vertical mirror lines OR half-turn rotation

ALL transformations can be used to map this pattern on to itself

pi (π)

π (a Greek letter spelt out as **pi**) is the symbol used to represent a particular number. It is an **irrational number** and the first 50 decimal places of its value are:

3.141 592 653 589 793 238 462 643 383 279 502 884 197 169 399 375 10 . . .
For the first 5000 places see the inside covers.
π relates the radius or diameter of a **circle** to its area and to its circumference.

π has a very long history, but it was not given the symbol and name that we use today until 1706 by the Welsh mathematician William Jones. However, not much notice was taken of his idea until it was published by the more famous Swiss mathematician Leonhard Euler in 1737.

The Ancient Egyptians (c.2000 BC) knew of the diameter/area relationship of a circle. They recorded that the area of a circle was found by taking eight-ninths of the diameter and squaring it. This implies a value of about 3.16 for π. The Babylonians, at about the same period, had a stated value of 3.125 which they used for their work on the circle.

Approximations for π (or its equivalent) that have been suggested at times are:

3 (*Bible: I Kings 7.23*)　　$\frac{22}{7}$ (*about 60 AD*)　　　　　$3\frac{17}{120}$ (*150 AD*)

3.1416 (*380 AD*)　　$\sqrt{10}$ (*600 AD*)　　$\frac{355}{113}$　　$\frac{333}{106}$　　$\sqrt{\sqrt{\frac{2143}{22}}}$

All of the dated estimates were produced by careful measuring and observation.

Archimedes was the first to work on the problem of finding the value of π in a systematic and analytical manner. He did it by 'squeezing' the circle between two similar polygons, one fitting tightly inside the circle and the other outside, and considering their areas.
The simplest case is:

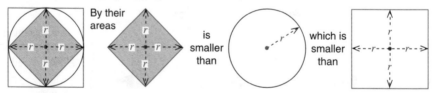

In modern terms: $2r^2 < \pi r^2 < 4r^2$ so $2 < \pi < 4$ (with a reasonable first guess for π of about 3). Archimedes started with a pair of hexagons and worked up to a pair of polygons having 96 edges. This allowed him to state that:

$$3\tfrac{10}{71} < \pi < 3\tfrac{1}{7}$$

which was, for those times, a remarkable feat of mathematics.

The polygon method was the only method for calculating π for many years. Only developments in arithmetic (especially the decimal system) allowed greater accuracy to be obtained with easier working. By 1630 it had been worked out to 39 decimal places, though this was obviously far in excess of anything needed for practical work. Such an 'accurate' value for π would give the circumference of the known Universe to an accuracy of less than the diameter of an atom.

The 17th century saw the beginnings of a whole new era of mathematics. Many great mathematicians lived at this time and generated much of the mathematics that we use today. One, among many, of the new ideas was that of a **series** and, in particular, the **Gregory series**, found by the Scottish mathematician John Gregory in 1671. It was:

$$\tan^{-1} x = x - \frac{x^3}{3} + \frac{x^5}{5} - \frac{x^7}{7} + \frac{x^9}{9} - \dots \quad (x \text{ is in radians})$$

which, provided x is less than 1 (and the smaller the better), will be accurate to several decimal places after a comparatively small number of terms. It can be used in conjunction with this formula to evaluate π:

$$\pi = 16 \tan^{-1}\left(\tfrac{1}{5}\right) - 4 \tan^{-1}\left(\tfrac{1}{239}\right)$$

The **Gregory series**, together with that formula, and others like it, was used from then on to calculate π. That method was used by William Shanks, who, after many years of calculating, finally published (in 1873) a value to 707 decimal places. That stood as the record until (in 1945) errors were found starting at the 528th place. These were corrected and the value then extended to 808 places. It was the last big calculation of π to be done 'by hand'.

In 1949 the first of the computer calculations of π was published, to 2037 places. It took 70 hours to compute using the same method as Shanks's. By 1967 it took only 28 hours to produce half a million digits using a faster series. And it didn't stop there. In 1997 a Japanese mathematician generated over 50 billion digits of π. It took 29 hours.

A new algorithm for computing π has been devised in recent years and is shown on the right. It starts with the values

$$A = 1, B = 0.5, C = \sqrt{2} \div 2, D = 0.25$$

It is very fast and gives a value for π accurate to 170 digits in only 6 loops, and to 1 million digits in 20 loops. Its big disadvantage is that, unlike the series method described above, all the accuracy needed must be worked with from the start, which makes impossible demands on the arithmetic processes of ordinary computers without special programs.

Start

Let $E = A$

Let $B = 2 \times B$

Let $A = (A + C) \div 2$

Let $C = \sqrt{C \times E}$

Let $D = D - [B \times (A - E)^2]$

Then $\pi = (A + C)^2 \div (4 \times D)$

Loop

Why? Why do they do it? Originally because it was there to be done, but there is now much more to it than that. It is a good standard exercise for checking the working of a computer. It presents an interesting challenge for those who have to write programs for computers. It provides a reason for the development of yet more mathematics. And there is always the quest to find some sort of 'pattern' in what looks to be a random set of digits. That quest has not ended, but neither has it yet produced anything of great significance.

polygon numbers

polygon numbers A polygon number is a number which states the quantity of objects needed for the objects to be arranged in the shape of a **regular polygon** with all the possible smaller similar polygons included in it. These polygons are made, starting with 1, as shown opposite. *The number is named after the shape, and a sequence can be formed of all the numbers which make that shape.*

figurate numbers ≡ **polygon numbers**

triangle numbers are **polygon numbers** having 3 edges.
The sequence begins 1, 3, 6, 10, 15, 21, 28, . . .
The nth triangle number is given by $n(n + 1) \div 2$

square numbers are those numbers that can be represented by the correct amount of dots laid out in rows and columns to make a square. *Some square numbers are:*

$$1 \bullet \qquad 4 \begin{smallmatrix} \bullet\ \bullet \\ \bullet\ \bullet \end{smallmatrix} \qquad 9 \begin{smallmatrix} \bullet\ \bullet\ \bullet \\ \bullet\ \bullet\ \bullet \\ \bullet\ \bullet\ \bullet \end{smallmatrix} \qquad 16 \begin{smallmatrix} \bullet\ \bullet\ \bullet\ \bullet \\ \bullet\ \bullet\ \bullet\ \bullet \\ \bullet\ \bullet\ \bullet\ \bullet \\ \bullet\ \bullet\ \bullet\ \bullet \end{smallmatrix}$$

$P_e(n)$ is used here to indicate a polygon number. It makes a polygon having e edges and that gives the number its name (3 = triangle, 4 = square etc.). n is the number of objects along the length of one edge, and is also the position of the number in the sequence.

		$n =$							
	name	1	2	3	4	5	6	7	8
$e = 3$	triangle	1	3	6	10	15	21	28	36
4	square	1	4	9	16	25	36	49	64
5	pentagon	1	5	12	22	35	51	70	92

Some values of $P_e(n)$ for various values of n and e

The general formula is $P_e(n) = n\,[\,2 + (e - 2)\,(n - 1)\,] \div 2$

centred-polygon numbers are those numbers made by taking e **triangle numbers** of the same size and adding 1. *As with the polygon numbers, their names are determined by the value of e. In making the actual shape, the 1 goes in the centre, and the triangles are arranged around it. Example: When $e = 4$ then a centred-square number can be made using any 4 triangle numbers (all the same size) plus 1.*

$C_e(n)$ is used here to indicate a centred-polygon number, using the definitions for n and e as given above. *The value of any centred-polygon number for given values of n and e can be found from the formula*

$$C_e(n) = [en(n - 1) \div 2] + 1$$

The study of polygon numbers goes back as far as 500 BC, but centred-polygon numbers were not devised until the 16th Century. The fascination has always been in discovering relationships between them (there are many) and inventing other types of 'shape-numbers'. There is no standardised notation for representing any of these numbers.

Polygon numbers and their growth: red shows what is being added each time.

triangle numbers
$P_3(1) = 1$ $P_3(2) = 3$ $P_3(3) = 6$ $P_3(4) = 10$ $P_3(5) = 15$

square numbers
$P_4(1) = 1$ $P_4(2) = 4$ $P_4(3) = 9$ $P_4(4) = 16$ $P_4(5) = 25$

hexagon numbers
$P_6(1) = 1$ $P_6(2) = 6$ $P_6(3) = 15$ $P_6(4) = 28$ $P_6(5) = 45$

Centred-polygon numbers and their growth

centred-triangle numbers:
$C_3(1) = 1$ $C_3(2) = 4$ $C_3(3) = 10$ $C_3(4) = 19$

centred-square numbers:
$C_4(1) = 1$ $C_4(2) = 5$ $C_4(3) = 13$ $C_4(4) = 25$

centred-hexagon numbers:
$C_6(1) = 1$ $C_6(2) = 7$ $C_6(3) = 19$ $C_6(4) = 37$ $C_6(5) = 61$

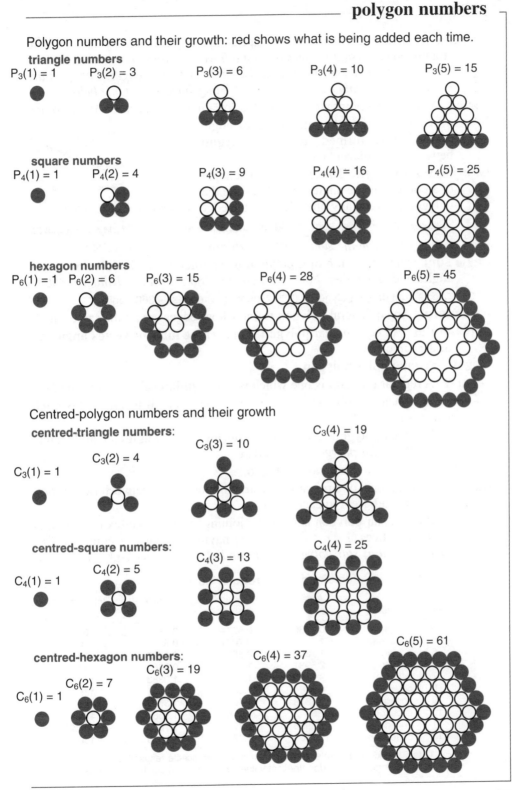

polygon A polygon is a plane *(= flat)* shape completely enclosed by three or more straight edges. *Usually edges are not allowed to cross one another, and the word is not often used for shapes having less than 5 edges. Polygons are named by the number of edges or angles they have – see table below.*

vertex A vertex is a point where two edges of a **polygon** meet to form a corner.

interior vertex angle An interior vertex angle is the angle formed inside a **polygon** between two adjacent edges.

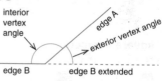

exterior vertex angle The exterior vertex angle of a polygon is the angle formed outside the polygon between any one edge and the edge adjacent to it, extended.

> Size of exterior vertex angle = 180° minus interior vertex angle *(it may be negative).*

For any polygon the sum of all the exterior vertex angles is 360°.

angle sum The angle sum of a **polygon** is the total of ALL its **interior vertex angles** added together. *For a triangle it is 180°; for a quadrilateral 360°*

> Angle sum of any polygon = (180 × number of edges) – 360 degrees

equilateral An equilateral **polygon** is one whose edges are all the same length.

equiangular An equiangular **polygon** is one whose **interior vertex angles** are all the same size.

isogon ≡ **equiangular polygon**

regular A regular **polygon** is one which is both **equilateral** and **equiangular**.

concave A concave **polygon** is one having at least one **interior vertex angle** which is greater than 180°.

convex A convex **polygon** is one whose **interior vertex angles** are all less than 180°. *All regular polygons are convex.*

re-entrant polygon ≡ **concave polygon**.

diagonals A diagonal of a polygon is a straight line drawn between two vertices that are not adjacent to each other.

star polygon A star polygon is made by joining every *r*th vertex of a polygon having *n* vertices $(1 < r < n - 1)$; *n* and *r* having no factors in common. *They are described as star polygons {n/r} and also known as n-grams.*

No. of edges	Name	Area = $e^2 \times \ldots$	C-radius = $e \times \ldots$	I-radius = $e \times \ldots$	Int. vertex angle °
3	triangle	0.4330	0.5774	0.2887	60
4	quadrilateral	1	0.7071	0.5	90
5	pentagon	1.7205	0.8507	0.6882	108
6	hexagon	2.5981	1	0.8660	120
7	heptagon	3.6339	1.1524	1.0383	128.57
8	octagon	4.8284	1.3066	1.2071	135
9	nonagon	6.1818	1.4619	1.3737	140
10	decagon	7.6942	1.6180	1.5388	144
11	undecagon	9.3656	1.7747	1.7028	147.27
12	dodecagon	11.196	1.9319	1.8660	150

Table of data for **regular** polygons

C-radius and I-radius refer to the circumcircle and incircle respectively.
e is the length of one edge. Inexact values are given to 5 significant figures.

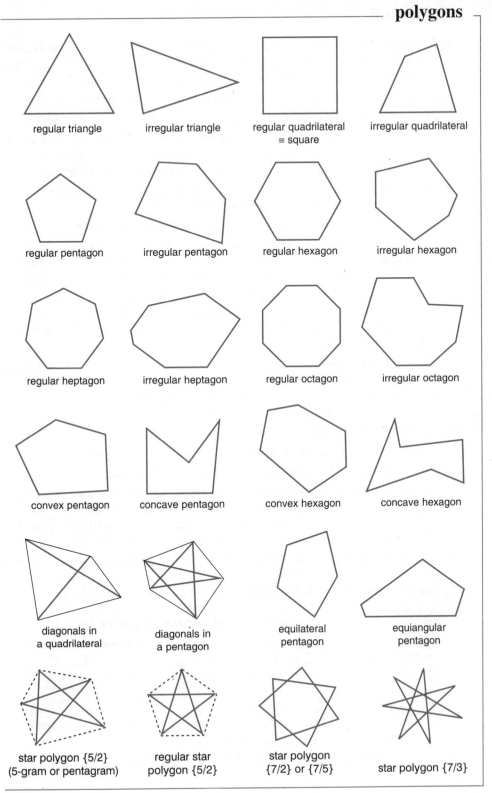

regular triangle | irregular triangle | regular quadrilateral ≡ square | irregular quadrilateral

regular pentagon | irregular pentagon | regular hexagon | irregular hexagon

regular heptagon | irregular heptagon | regular octagon | irregular octagon

convex pentagon | concave pentagon | convex hexagon | concave hexagon

diagonals in a quadrilateral | diagonals in a pentagon | equilateral pentagon | equiangular pentagon

star polygon {5/2} (5-gram or pentagram) | regular star polygon {5/2} | star polygon {7/2} or {7/5} | star polygon {7/3}

polyhedrons

polyhedron A polyhedron is a 3-dimensional shape whose faces are all **polygons**. *It must have at least 4 faces. Its name is based on the number of faces it has. Example: a tetrahedron has 4 faces, a pentahedron has 5 faces, and so on.*

edge An edge is a straight line formed where 2 faces of a **polyhedron** meet.

vertex A vertex of a **polyhedron** is the angular point where 3 or more **edges** meet.

net A net is an arrangement of polygons connected at their edges, all lying in one plane *(= flat surface)* which can be folded up to make a **polyhedron**. *There is always more than one way of doing this. It is also known as a 'development'.*

convex A convex polyhedron is one in which any straight line joining one **vertex** to another lies entirely on or inside the polyhedron.

non-convex A non-convex polyhedron is one which is not **convex**.

regular A regular polyhedron has all of its faces identical, and the same number of **edges** meeting at each **vertex**. *There are only 9 possibilities, of which 5 are convex. These 5 are known as the* **Platonic solids**. *See table below.*

semi-regular A semi-regular polyhedron is a **polyhedron** of which every face is a **regular polygon** and every **vertex** is identical. *Excluding prisms and antiprisms, there are only 13 possibilities and these are known as the* **Archimedean solids**.

deltahedron A deltahedron is a **polyhedron** whose every face is an **equilateral triangle**. *There are only 8 possibilities of the convex type. Three of these are the regular tetra-, octa- and icosa- hedrons. The other five are irregular and have 6, 10, 12, 14 or 16 faces.*

hexahedron A hexahedron is a **polyhedron** having 6 faces.

cube A cube is a regular **hexahedron** and all its faces are squares.

cuboid A cuboid is a **hexahedron** whose faces are all rectangles.

circum-sphere A circum-sphere is the sphere which can be drawn around the OUTSIDE of a **polyhedron** so as to pass through ALL its vertices. *It is always possible to draw a circum-sphere for a regular convex polyhedron, and for any* **Archimedean solid**, *but it may not be possible for others.*

in-sphere An in-sphere is the sphere which can be drawn INSIDE a **polyhedron** so as to touch ALL its faces. *It is always possible to draw an in-sphere for a regular convex polyhedron, but it may not be possible for others.*

No. of faces	Table of data for **regular convex polyhedrons**				
	Name	Area = $e^2 \times ...$	Volume = $e^3 \times ...$	C-radius = $e \times ...$	I-radius = $e \times ...$
4	tetrahedron	1.73205	0.117851	0.612372	0.204124
6	cube	6	1	0.866025	0.5
8	octahedron	3.46410	0.471405	0.707107	0.408248
12	dodecahedron	20.6458	7.66312	1.40126	1.113516
20	icosahedron	8.66025	2.18170	0.951057	0.755761

Area is total surface area of the polyhedron.
C-radius and I-radius refer to the circum-sphere and in-sphere respectively.
e is the length of one edge. Inexact values are given to 6 significant figures.

The 5 regular convex polyhedrons and some nets

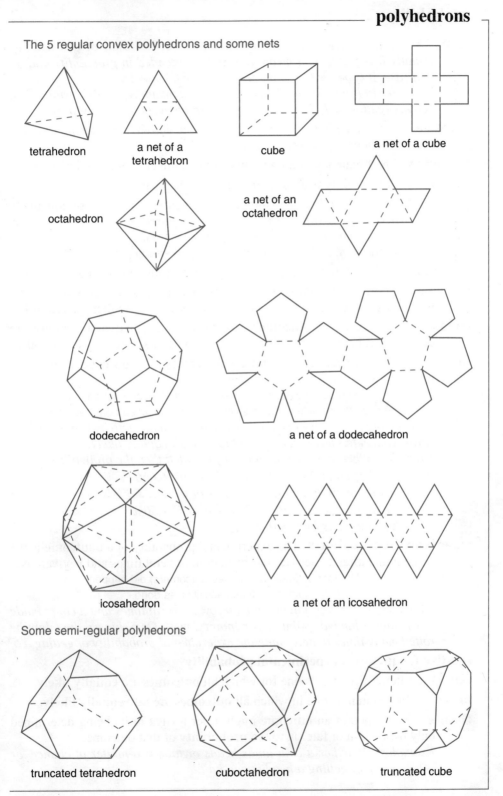

tetrahedron

a net of a
tetrahedron

cube

a net of a cube

octahedron

a net of an
octahedron

dodecahedron

a net of a dodecahedron

icosahedron

a net of an icosahedron

Some semi-regular polyhedrons

truncated tetrahedron

cuboctahedron

truncated cube

event An event is EITHER an activity OR some specific result of that activity. *Usually it is the second meaning which is intended in probability and, in that case, it is better to use the word 'outcome'.*
Example: Rolling a single die is an event (1st meaning), and a result of that activity (getting a 1, 2, 3, 4, 5 or 6) is also an event (2nd meaning).

outcome An outcome is the actual result of some activity. *It is identical to the second meaning of* **event** *given above.*

frequency The frequency of an **outcome** is the number of times it happens.

possibility A possibility is an **outcome** that CAN happen.

probability The probability of an **outcome** (or event) is a measure of how likely that outcome is. *A value of 0 means it is impossible, while 1 means it is certain; otherwise the value must lie between 0 and 1. It may be given as a common fraction, a decimal fraction or a percentage.*

P() or Pr() is the symbol for the **probability** of the **outcome** named in brackets.

probability scale A probability scale is a line numbered 0 to 1 or 0 to 100% on which **outcomes** (or events) can be placed according to their **probability**.

equally likely A set of **outcomes** associated with a particular activity are described as being equally likely when each occurs as readily as any other.

theoretical probability The theoretical probability of an **outcome** is the value predicted from the fraction given by:

$$\frac{\text{Number of ways that named outcome(s) can happen}}{\text{Number of all possible outcomes which can be obtained from that activity}}$$

This can only be done when the item(s) on which the activity is based (dice, cards, coins etc.) have outcomes which are all equally likely.
Examples: When a normal die (1 to 6) is rolled then the probability of:
getting a 4 is $P(4) = \frac{1}{6}$ or 0.166666 or 16.666 %
getting more than 4 is $P(>4) = \frac{2}{6}$ or $0.\dot{3}$ or 33.3%
getting 1, 2 or 5 is $P(1,2,5) = \frac{3}{6}$ or 0.5 or 50%
getting a 7 is $P(7) = 0$

experimental probability The experimental probability of an **outcome** is the value found after an activity has been done several times and is given by:

$$\frac{\text{Number of times that named outcome(s) did happen}}{\text{Number of times activity was done}}$$

Examples: A drawing-pin might be thrown several times and a count made of whether it landed 'point up' or 'point down'. Or a biased die might be rolled many times to determine the experimental probability of getting a 5

relative frequency ≡ **experimental probability**

fair A fair item (dice etc.) is one for which all **outcomes** are **equally likely**.

biased A biased item is one in which all **outcomes** are NOT **equally likely**.

chance The chance of an **outcome** might refer EITHER to its being determined only by a whim of fate OR to the **probability** of that outcome.
Examples: 'Whether a coin lands heads or tails is a matter of chance' or 'The chance of getting a head is one-half.'

mutually exclusive events are sets of **events** or **outcomes** for which the happening of one of them means that none of the others can happen.
Example: When rolling a die the outcomes are mutually exclusive, since when one number comes to the top it must mean that none of the others can.

independent events Two or more **events** or **outcomes** are independent if the happening of one of them has no effect on the other.
Example: When two dice are rolled there are two independent outcomes, since the number showing on one does not influence the number on the other.

dependent events Two **events** or **outcomes** are dependent if a statement or probability for one of them affects a statement or probability for the other.
Example: One box holds 4 red and 6 black marbles; another holds 1 red and 9 black marbles. The probability of choosing a red marble from one box must depend on which box is chosen.

combined events describe the putting together of two or more separate **events** or **outcomes** to be considered as one single event or outcome. *This is usually done in order to find the probability of a final single outcome. The separate outcomes might be independent of, or dependent upon, each other. Examples: Rolling 2 dice (or 1 die twice) and adding the separate scores is combining 2 independent outcomes. Taking 2 counters from a bag of mixed colours WITHOUT replacing the first is combining 2 dependent outcomes.*

compound events ≡ **combined events**

conditional probability is the **probability** of an **outcome** happening when it is **dependent** upon, or following, some other outcome.
Example: A bag contains 8 red and 2 black counters. The probability of drawing 2 red counters, if the first drawn is not replaced, is given by the probability of the first counter being red times the probability of the second being red, which is $\frac{8}{10} \times \frac{7}{9} = \frac{28}{45}$

tree diagrams are drawn to find and display all possible results when several outcomes are being combined.
Example: When 3 coins are tossed all possible results can be found and displayed by using a tree diagram like that on the right (H=Head, T=Tail).

$$H = HHH = 3H$$
$$T = HHT = 2H + T$$
$$H = HTH = 2H + T$$
$$T = HTT = H + 2T$$
$$H = THH = 2H + T$$
$$T = THT = H + 2T$$
$$H = TTH = H + 2T$$
$$T = TTT = 3T$$

Start

odds are another type of **probability** and the odds against a successful **outcome** happening are given by:
number (of *other* outcomes in the activity) TO number (of ways outcome can happen)
Example: The odds against getting a '3' with a single die are 5 to 1 since there are 5 other numbers and only one '3', so there are 5 ways of losing against only 1 way of winning. The probability of getting a 3 is $\frac{1}{6}$ or $\frac{1}{5+1}$

Odds of a to b change to a probability of $\dfrac{b}{(a + b)}$

A probability of $\frac{a}{b}$ changes to odds of $(b - a)$ to a

evens When the **odds** are 1 to 1 they are even. *The probability for evens is $\frac{1}{2}$*

pyramid A pyramid is a **polyhedron** having any polygon as one face with all the other faces being triangles meeting at a common vertex. *The pyramid is named after the polygon forming the face from which the triangles start.*

base The base of a **pyramid** is the polygonal face which names the pyramid.

apex The apex of a **pyramid** is the vertex at which the triangular faces meet.

perpendicular height The perpendicular height of a **pyramid** is the distance of its **apex** from the plane of its **base**.

> Volume of pyramid = Area of base × Perpendicular height ÷ 3

altitude ≡ **perpendicular height**

vertex In the case of a **pyramid**, vertex is often used to mean the **apex**.

right pyramid A right pyramid is one having all its triangular faces equal in size. *The base is a regular polygon, the apex is perpendicularly above the centre of the base, and all the triangular faces make the same angle with the base.*

right square-based pyramid A right square-based pyramid is a **right pyramid** having a square base. *It is what is usually meant when only the word 'pyramid' is used and is the type seen in Egypt as a tomb of the Pharaohs.*

oblique pyramid An oblique pyramid is a NON-**right pyramid**.

slant height The slant height of a **pyramid** is the length of a perpendicular from the mid-point of a base-edge to the apex. *The slant heights of a right pyramid are all the same length.*

slant edge The slant edges of a **pyramid** are all those edges joined to the **apex**. *The slant edges of a right pyramid are all the same length.*

frustum of a pyramid A frustum of a pyramid is the part of a **pyramid** cut off between the **base** and a plane which is parallel to the base.

> Volume of frustum $= (A + B + \sqrt{AB}) \times h \div 3$ where
> A, B = areas of top and bottom faces of frustum
> h = distance between faces

cross-section A cross-section of any 3-D shape is the 2-D figure shown when that shape is cut across, in some specified place and direction, by a plane.

prism A prism is a **polyhedron** having 2 faces identical and parallel to each other (usually referred to as the 'ends' or 'bases'), and any plane cut made parallel to the ends produces a cross-section the same shape and size as the ends. *All faces, other than the ends, are rectangles or parallelograms. Prisms are named after the shape of the cross-section (if it has a name) as in 'triangular prism' or 'hexagonal prism'. If the other faces are rectangles, it is also referred to as a* **right prism**.

> The volume of a *prism* can be found by multiplying the area of one of the end faces by the perpendicular distance between the two ends.

antiprism An antiprism is a **polyhedron** that has 2 faces identical and parallel to each other. All the other faces are identical triangles, with each vertex of every triangle touching a vertex of one of the end faces, so that 4 edges meet at every vertex. *Unlike with a prism, its cross-section varies.*

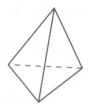

triangle-based pyramid
≡ tetrahedron

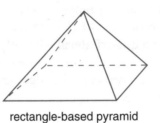

rectangle-based pyramid

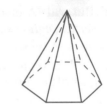

hexagon-based pyramid

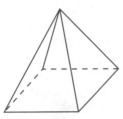

right square-based
pyramid

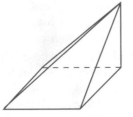
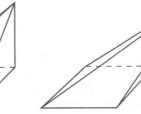

oblique pyramids

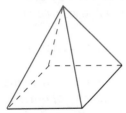

slant edges

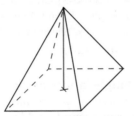

perpendicular height

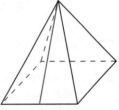

slant height

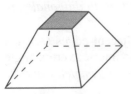

frustum of a right
square-based pyramid

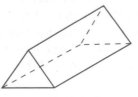

triangular prism

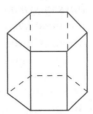

hexagonal prism

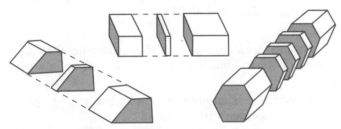

some prisms and their cross-sections

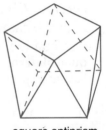

square antiprism

quadrilateral A quadrilateral is a **polygon** which has 4 edges. *Its 4 interior vertex angles (=corners) add up to 360 degrees.*

trapezium A trapezium is a **quadrilateral** with only one pair of parallel edges.

trapezoid ≡ **trapezium** in N. American usage, but in UK usage it is a quadrilateral in which no two opposite edges are parallel.

isosceles trapezium An isosceles trapezium is a **trapezium** in which the two opposite edges, which are not parallel, are the same length. *It has one line of symmetry and both diagonals are the same length.*

parallelogram A parallelogram is a **quadrilateral** which has two pairs of parallel edges. *It has rotational symmetry of order 2, and its diagonals bisect each other. Usually one pair of edges is longer than the other pair, no interior vertex angle (=corner) is a right angle and it has no lines of symmetry.*

rhombus A rhombus is a **quadrilateral** whose edges are all the same length. *Its diagonals bisect each other at right angles and both are also lines of symmetry. Usually no interior vertex angle (=corner) is a right angle and then it is sometimes referred to as a* **diamond, lozenge,** *or* **rhomb.**

rhomboid A rhomboid is a **parallelogram** having adjacent edges of different lengths. *The word is little used because of possible confusion.*

> The area of a trapezium, parallelogram or rhombus can be found by adding together the lengths of one pair of parallel edges, dividing by 2, and multiplying this by the perpendicular distance between them.

rectangle A rectangle is a **quadrilateral** in which every interior vertex angle *(= corner)* is a right angle.

oblong An oblong is a **rectangle** in which one pair of edges is longer than the other pair. *It has two lines of symmetry and rotational symmetry of order 2. Both diagonals are the same length and bisect each other.*

square A square is a **rectangle** whose edges are all the same length. *It has four lines of symmetry and rotational symmetry of order 4. Both diagonals are the same length and bisect each other at right angles.*

kite A kite is a **quadrilateral** which has two pairs of adjacent edges *(= edges which are next to each other)* of the same length, and no interior vertex angle *(= corner)* is bigger than 180 degrees. *It has one line of symmetry and its diagonals cross each other at right angles.*

arrowhead An arrowhead is a **quadrilateral** which has two pairs of adjacent edges of the same length and ONE interior vertex angle *(= corner)* which is bigger than 180 degrees. *It has one line of symmetry and its diagonals do not cross. It is also known as a* **dart** *or* **deltoid.**

regular quadrilateral ≡ **square**.

irregular quadrilateral Strictly speaking, an irregular quadrilateral is any **quadrilateral** that is not a square, but it is usually taken to be one not having a special name.

golden rectangle A golden rectangle is an **oblong** with its two edge-lengths sized in the proportions of the **golden ratio** (≈ 1.618 : 1).

> Length of longer edge ≈ 1.618 × length of shorter edge.

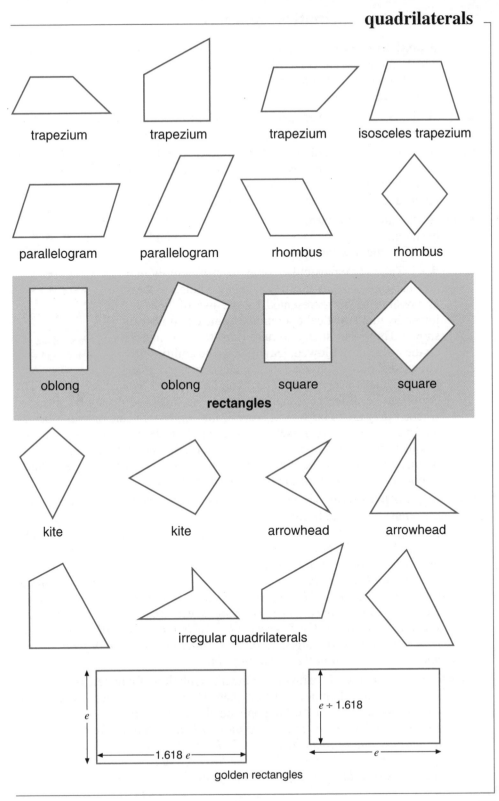

trapezium   trapezium   trapezium   isosceles trapezium

parallelogram   parallelogram   rhombus   rhombus

oblong   oblong   square   square

rectangles

kite   kite   arrowhead   arrowhead

irregular quadrilaterals

e $e \div 1.618$

1.618 e e

golden rectangles

recreational mathematics covers games, puzzles and similar activities in which mathematical principles are used in some way: to create them, to play them, or to solve them. The types and examples given here are just a sample drawn from the very wide range of material that is available.

cryptarithms are sums in arithmetic in which some (or all) of the original digits have been hidden in some way, and it is required to find them.

alphametics are **cryptarithms** in which all of the digits have been replaced by letters (each letter always representing the same digit, and each digit always represented by the same letter) and with the arrangement forming real words.

```
  S E N D
+ M O R E
---------
M O N E Y
```

asterithms are **cryptarithms** in which some of the digits have been replaced by asterisks, and the puzzle is to find the correct values of the missing digits.

```
  * 4 6
+ 2 8 *
-------
  4 * 1
```

four 4's This is a particular type of problem which originated in 1871. It requires as many whole numbers as possible to be represented by the use of four 4's joined by mathematical operations. Some examples are shown. There are usually no rules or restrictions placed on the operations allowed, leaving it to the abilities and knowledge of the individual. From this beginning other

$1 = (4 + 4) \div (4 + 4)$
$2 = (4 \div 4) + (4 \div 4)$
$3 = (4 + 4 + 4) \div 4$
$4 = 4 + 4 - \sqrt{4} - \sqrt{4}$
$15 = 44 \div 4 + 4$
$17 = 4 \times 4 + 4 \div 4$
$27 = 4! + \sqrt{4} + (4 \div 4)$

problems have been posed. The 4 can be replaced by any other digit. The digits can be different from each other – in this case it may be specified whether they are to be kept in order or not. In recent years there has been a tendency to use the year itself, so for example 1998 yields $1 \times 9 + 9 - 8 = 10$. Another variation is to ask for the largest (or smallest) possible number to be made using only the digits given. In all of this work it is particularly important to be aware of the **order of operations**.

counting-out problems are based on the idea that a set of objects (or people) are arranged in a circle and then, after counting up to some number n while moving around the circle, the nth object is removed; this being continued until only one object is left. *It is often seen in children's games where one has to be chosen for a certain task, and the counting is done by using a rhyme. Problems are based on a requirement to predetermine where in the circle an object should be placed to ensure it is selected.*
Example: 12 playing cards (Ace to King) are arranged in a circle and then, starting the count at the 5, every 7th card is removed until none are left. How must they be arranged so that the cards removed are in their correct numerical order, from the Ace to the King?

tricks There are many tricks involving the manipulation of numbers (often using cards, dice and dominoes) which seem to produce 'magical' results, but which depend for their working only on elementary principles of arithmetic. *Example: Ask someone to roll two dice without revealing the results. Then to multiply one of them by 5, add 7, double the result and add the other number, and give the result. From the two digit answer, subtract 14 and the two digits remaining will give the numbers showing on the dice.*

measuring problems have always been popular because they are easy to understand, and do relate to something familiar in ordinary life, though some constraints have to be introduced in order to make it a problem. One of the earliest type was introduced by **Tartaglia**: *There are three uncalibrated containers of 3, 5 and 8 litres capacity and the largest one is full. How can the 8 litres be divided into two lots of 4 litres?*
Some measuring problems are based on the use of weights, while others require lengths to be measured with imperfectly marked rulers.

ferry problems is another class of problems which are usually easy to understand but not so easy to resolve. The oldest (and easiest) of this type concerns the traveller with a wolf, a goat and a bag of cabbages who needs to cross a river using one boat and taking only one of the things at a time. But, the wolf and goat can never be left alone together, and nor can the goat and cabbages.

Diophantine problems seem at first sight to have insufficient information for their solution but, because the objects involved can only be counted using whole numbers (such as live animals) at least one solution can be found.
Example: In the market, ducks cost £5 each, hens cost £1 each and baby chickens were 20 for £1. Kim bought at least one of each, a 100 birds altogether, for a total of £100. How many of each did Kim buy?

magic squares A magic square is a set of numbers arranged in the form of a square so that the total of every **row**, **column** and **diagonal** is the same. *It is usual to require that every number is different. In most cases the numbers also form some kind of sequence. Examples:*

using 1 to 9			using 1 to 16				using primes only		
8	3	4	16	2	3	13	17	89	71
1	5	9	5	11	10	8	113	59	5
6	7	2	9	7	6	12	47	29	101
			4	14	15	1			

Latin squares A Latin square of size *n* by *n* is one in which *n* different objects are each repeated *n* times and arranged in a square array so that none of them is repeated in any row or column.

Graeco-Latin squares A Graeco-Latin square is made by combining two different **Latin squares** (each of which is made of a different set of objects) so that no pair of objects is repeated. It is also known as an **Euler square**. *Squares of this type are useful in the design of experiments. Example:*

A	B	C	D		1	2	3	4		A1	B2	C3	D4
C	D	A	B	**+**	4	3	2	1	**=**	C4	D3	A2	B1
D	C	B	A		2	1	4	3		D2	C1	B4	A3
B	A	D	C		3	4	1	2		B3	A4	D1	C2

A problem based on this is to take 16 playing cards, A, K, Q, J of each suit, and put them in a 4 by 4 array, so that no two cards of the same value or suit are in line.

recreational mathematics (spatial)

dissections Dissection puzzles require one shape to be cut up into a number of pieces and these pieces then re-assembled to make some other shape. *In some cases the pieces are given and the puzzle is to find how they can be used to make another shape – which is usually specified.*
Example: Dissect an equilateral triangle into the smallest possible number of pieces that can be rearranged to make a square.

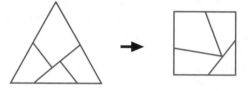

tangram Tangram puzzles use a standard **dissection** of a square into 7 pieces and require arrangements of those pieces to be found which will make up a given shape. *The given shape is shown only in outline.*

Making the pieces

The grid is only for guidance. The red lines show how the square is cut up.

A tangram puzzle and its solution

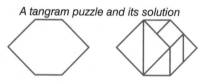

polygrams is the general name for those puzzles which are similar in type to tangrams, but which use a different dissection of the square, or even a different shape altogether. *Below are some varieties that have been commercially produced in the past, with their names.*

| Pythagoras | Chie No Ita | Tormentor | Cross Breaker |

polyiamonds are 2-dimensional shapes made from identical equilateral triangles joined together by their edges. *Each type is named from the number of triangles it uses: 4 triangles = tetriamond; 5 triangles = pentiamond etc.*

hexiamonds are the 12 different **polyiamonds** that can be made from 6 equilateral triangles. *They are shown below, with their names.*

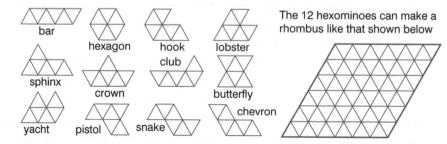

bar, hexagon, hook, club, lobster, sphinx, crown, butterfly, yacht, pistol, snake, chevron

The 12 hexominoes can make a rhombus like that shown below

polyhexes are 2-dimensional shapes made from identical regular hexagons.

polyominoes are 2-dimensional shapes made from identical squares joined together by their edges. *Each type takes its particular name from the number of squares it uses: 4 squares = tetromino; 6 squares = hexomino etc. The number of different shapes that can be made with any given number of squares can be worked out, but there is no formula to determine this relationship, which is given in the following table:*

No.of squares	1	2	3	4	5	6	7	8	9	10
No. of shapes	1	1	2	5	12	35	108	369	1285	4655

pentominoes are the 12 different **polyominoes** that can be made from 5 squares. *Of all the polyominoes, these have proved to be the most popular set. The set is big enough to work with, but small enough to be handleable. Each piece is identified by a letter as shown below. One puzzle (out of many) for these pieces is to make a rectangle using all 12 of them.*

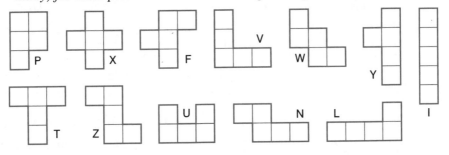

polycubes are 3-dimensional shapes made from identical cubes joined together by their faces. *The number of different shapes that can be made with any given number of cubes is always of interest, but there is no formula to determine this relationship, which is given in the following table:*

No.of cubes	1	2	3	4	5	6	7	8	9	10
No. of shapes	1	1	2	8	29	166	1023	6922	48311	346543

Soma cubes are the 7 different **polycubes** that can be made from 3 or 4 cubes, with each shape having at least one concave edge. *It needs 27 cubes to make all seven. They can be assembled to make models of a wide variety of objects, and can be used to make a cube in 240 different ways.*

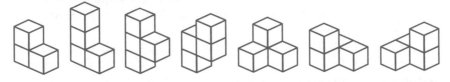

illusions An illusion is a visual trick showing something that cannot exist. The trick on the right shows an area of 64 squares cut up and rearranged to make a 5 by 13 oblong with an area of 65 squares.

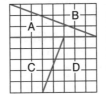

sequences and series

sequence A sequence is a set of numbers or objects made and written in order according to some mathematical rule.

term A term is one of the separate numbers or objects of a **sequence**. *The terms of a sequence are usually separated by a comma and a space.*

random sequence A random sequence is a set of numbers or objects made and written in order, according to NO apparent rule and for which, no matter how many **terms** are known, the next cannot be predicted with certainty.

natural numbers The natural number sequence is that **sequence** which is used for counting.
The sequence begins (10 terms shown): 1, 2, 3, 4, 5, 6, 7, 8, 9, 10, …

doubling sequence A doubling sequence is a **sequence** in which each term is twice (× 2) the value of the previous term. *It usually starts with 1*
The sequence begins (10 terms shown): 1, 2, 4, 8, 16, 32, 64, 128, 256, 512, …

lucky number sequence The lucky number sequence is made from the **natural numbers** by first deleting every 2nd number; from those that are left delete every 3rd number; from those delete every 4th number; then every 5th , 6th, 7th and so on until no more can be deleted; those remaining form the sequence. *Rules can be made up for generating all sorts of sequences.*
The sequence begins (10 terms shown): 1, 3, 7, 13, 19, 27, 39, 49, 63, 79, …

recursive sequence A recursive sequence is a **sequence** in which each new **term** is defined in relation to some terms which have been made previously.

Fibonacci sequence The Fibonacci sequence is a **recursive sequence** where, starting with the first two terms as 1, 1, each new term is made by adding together the two previous terms.
The sequence begins (10 terms shown): 1, 1, 2, 3, 5, 8, 13, 21, 34, 55, …
Formally this is written as $F_n = F_{n-2} + F_{n-1}$ where $F_1 = 1$ and $F_2 = 1$

The value of the nth term can be found from the formula:

$$F_n = \frac{1}{\sqrt{5}} \left(\frac{1 + \sqrt{5}}{2} \right)^n$$ *rounded to the nearest whole number*

Lucas sequence The Lucas sequence is a **recursive sequence** where, starting with the first two terms as 1, 3, each new term is made by adding together the two previous terms.
The sequence begins (10 terms shown): 1, 3, 4, 7, 11, 18, 29, 42, 71, 113, …
Formally this is written as $L_n = L_{n-2} + L_{n-1}$ where $L_1 = 1$ and $L_2 = 3$

arithmetic progression or **AP** An AP is a **sequence** where each new **term** after the first is made by ADDING on a constant amount to the previous term.
Example: 3, 7, 11, 15, 19 , … with a first term of 3 and a constant of 4.

geometric progression or **GP** A GP is a **sequence** where each new **term** after the first is made by MULTIPLYING the previous term by a constant amount.
Example: 2, 6, 18, 54, 162, … with a first term of 2 and a constant of 3 or: 14, 7 3.5, 1.75, 0.875, … with a first term of 14 and a constant of 0.5

series A series is made by placing addition (or subtraction) signs between the **terms** of a **sequence**.

arithmetic series An arithmetic series is an **AP** with addition (or subtraction) signs inserted between the various terms.

Example: Using the previous AP, the series is 3 + 7 + 11 + 15 + 19 + . . .

To find the sum (= total) of an arithmetic series over n terms use the formula

$$na + n(n-1)d \div 2 \qquad \text{where}$$

a is the value of the first term

n is the number of terms

d is the constant added to each term

geometric series A geometric series is a **GP** with addition (or subtraction) signs inserted between the various terms.

To find the sum (= total) of a geometric series over n terms use the formula

$$\frac{a(r^n - 1)}{r - 1} \qquad \text{where}$$

a is the value of the first term

n is the number of terms

r is the constant multiplier ($r \neq 1$)

infinite series An infinite series is a **series** whose terms are never ending. *There is no last term; it is always possible to write another one.*

convergent series A convergent series is an **infinite series** which, as more and more terms are used, moves towards some definite value. *An infinite* **geometric series** *is convergent if* $|r| < 1$, *otherwise it is* **divergent**.

divergent series A divergent series is an **infinite series** which is not **convergent**. *Usually it means the value of the series grows as more terms are used. An infinite* **geometric series** *is divergent if* $|r| \geq 1$

Σ is the symbol (known as a summation sign) used to show that a **series** is being given. *The symbol is followed by an expression showing how each term in the series is made.*

Example: $\Sigma \frac{1}{n}$ *means* $\frac{1}{1} + \frac{1}{2} + \frac{1}{3} + \frac{1}{4} + \frac{1}{5} + \frac{1}{6} + \frac{1}{7} + \ldots$

which is known as the **harmonic series**. *It is* **infinite** *and* **divergent**.

alternating series An alternating series is an **infinite series** in which the signs between the terms are alternately for addition (+) and subtraction(−).

Example: $\frac{1}{1} - \frac{1}{2} + \frac{1}{3} - \frac{1}{4} + \frac{1}{5} - \frac{1}{6} + \frac{1}{7} - \frac{1}{8} + \ldots$

which, in spite of its resemblance to the **harmonic series**, *is* **convergent**.

partial sum A partial sum of an **infinite series** is found by using only a finite number of terms. *The starting and finishing terms to be used are indicated by the* **n** *values written respectively below and above the summation sign.*

Example: $\sum_{n=1}^{n=7} n$ *means* $1 + 2 + 3 + 4 + 5 + 6 + 7 \ (=28)$

continued product A continued product is made by placing multiplication signs between the **terms** of a **sequence**.

Π is the symbol used to show that a **continued product** is being given.

Example: $\prod_{n=1}^{n=7} n$ *means* $1 \times 2 \times 3 \times 4 \times 5 \times 6 \times 7 \ (=5040)$ *and is another way of writing* **factorial n** *or* **n!** *or, in this case, 7!*

set A set is a collection of objects (letters, numbers or symbols etc.) which is defined EITHER by listing all the objects OR by giving a rule that allows a decision to be made as to whether or not an object belongs in that set. *Sets are usually shown listed or defined within curly brackets:{ }.*

Examples: {a, e, i, o, u} is a listed set that could also be described by the rule {the vowels}

{5, a person, a table, Z} is a listed set for which a rule would be difficult to find.

{all the numbers} is a rule for a set that it is impossible to list.

universal set A universal set is a **set** which is first defined (by list or rule) and within which all the statements that follow must be interpreted.

Examples: After giving the universal set {positive numbers less than 10} then the set {even numbers} would be only {0, 2, 4, 6, 8}

In the universal set {all positive numbers} $x^2 = 4$ has only the solution $x = 2$ since $^-2$ is not in the universal set.

$\mathcal{E}$ is the symbol for the **universal set**, but others are used.

universe ≡ **universal set**

member A member of a **set** is one of the objects contained in that set.

element ≡ **member**

∈ is the symbol meaning 'is a member of' or 'belongs to'.

Example: $2 \in$ {even numbers}

empty set The empty set is the **set** which has NO **members**.

Example: The set {all odd numbers divisible by 2} is empty.

∅ is the symbol for the **empty set**.

null set ≡ **empty set**

subset A subset is a **set** which contains part of (or all of) another set.

⊂ is the symbol meaning 'is a subset of'.

Example: $\{2, 7, f, t, M, \phi\} \subset$ {numbers, letters, symbols}

proper subset A proper subset is a **subset** which does NOT contain ALL the members of the other set and is not ∅.

complement The complement of a **set** is all those **members** which are NOT in that set, but which ARE in the **universal set** originally given.

′   is the symbol for the complement of a **set.**

Example: Suppose the universal set is {odd numbers less than 30} and the set A is {all prime numbers}.

Then the complement of A is shown by A′ and is {1, 9, 15, 21, 25, 27}

union The union of 2 (or more) **sets** is their combination into a single set containing ALL the **members** of the original sets. *A member found in more than one of the original sets need only be shown once in the union.*

∪ is the symbol for the **union** of sets.

Example: $\{4, 7, 13, 20\} \cup \{2, 7, 10\}$ is $\{2, 4, 7, 10, 13, 20\}$

intersection The intersection of 2 (or more) **sets** is the single set made containing ONLY **members** which are common to ALL the original sets.

∩ is the symbol for the **intersection** of sets.
  Example: *{4, 7, 13, 20} ∩ {2, 7, 10} is { 7}*

disjoint Disjoint **sets** are those having NO **members** in common.

superset If A is a **subset** of B, then B is said to be a superset of A.

⊃ is the symbol meaning 'includes' or 'is a superset of'.
  Example: A ⊃ B can be read as 'set A includes set B.' or 'set A is a superset of set B' or 'set B is a subset of set A'.

symmetric difference The symmetric difference of 2 **sets** is the single set made which contains only **members** which were found ONLY ONCE in the original sets. *A member found in both sets would not be included.*

∇ is the symbol for the **symmetric difference** of two sets.
  Example: *{2, 5, 8, 12} ∇ {1, 5, 12, 15} is { 1, 2, 8, 15}*

Venn diagrams are used to give a pictorial view of the relationships of **sets** and **subsets** within a **universal set**; the universal set is shown enclosed by a rectangle, and all others by circles or simple closed curves.
  Examples:

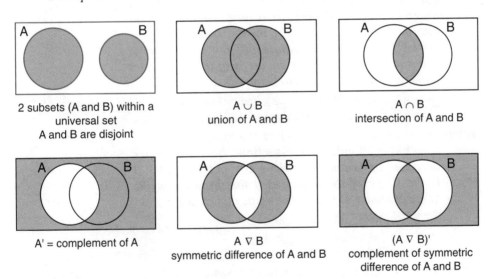

2 subsets (A and B) within a universal set
A and B are disjoint

A ∪ B
union of A and B

A ∩ B
intersection of A and B

A' = complement of A

A ∇ B
symmetric difference of A and B

(A ∇ B)'
complement of symmetric difference of A and B

finite set A finite set is a **set** whose **members** can be counted.

infinite set An infinite set is a **set** whose **members** cannot be counted and the quantity stated in terms of any defined number.
  Example: The set of all real numbers is an infinite set.

enumerate To enumerate a **set** (or **subset**) is to list all its **members**.

denumerable A denumerable **set** is one for which a mapping can be established which puts all its members into a one-to-one correspondence with the positive integers.

parallelepiped A parallelepiped is a **solid** having 6 faces, each of which is a parallelogram. *Opposite faces are parallel. A* **cuboid** *is a special case of a parallelepiped where each of the faces is a rectangle.*

> Volume = Area of 1 face mulltiplied by the perpendicular distance between *that* face and the opposite face.

rhombohedron A rhombohedron is a **parallelepiped** whose faces are rhombuses. *In scientific work (on crystals) it is often called a* **rhomboid**.

barrel A barrel, of the traditional type and made of wood, is roughly cylindrical in shape having two circular ends of the same diameter, but bulging outwards in the middle of its height (or length) to a bigger diameter.

> It is difficult to calculate the capacity of a barrel exactly, but a very good approximation can be found using:
>
> $$\frac{\pi h}{360}(39D^2 + 26Dd + 25d^2)$$ where d = diameter of one end
> D = diameter in middle
> h = height of barrel

lune When two circles (lying in the same plane) overlap, then a lune is the shape which is formed between the outside of one circle and the inside of the other. *In the diagram, the shaded area is a lune. The two circles may be the same, or different, in size. Two overlapping circles form two lunes whose difference in areas is equal to the difference in the areas of the two circles making them. A lune is also known as a* **crescent**.

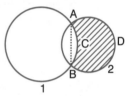

> The area of lune ADBC can be found by subtracting the area of segment ABC of circle 1 from the area of segment ABD of circle 2.

torus A torus is a solid ring, in which the 'band' making the ring has a circular **cross-section**. *It is sometimes described as 'a doughnut with a hole in the middle'. It is the shape of the object used for throwing in the game of quoits. It is also known as an* **anchor-ring**.

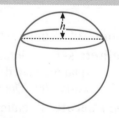

> Volume $=\frac{\pi^2 d^2}{4}(D + d)$ where D = inside diameter of ring
>
> Surface area $= \pi^2 d(D + d)$ d = diameter of circular cross-section

segment of a sphere A segment of a **sphere** is the shape cut off by a single plane which passes through the sphere. *The plane divides the sphere into 2 segments. A plane cutting through the centre of the sphere produces two identical segments which are* **hemispheres**. *A segment which is smaller than a hemisphere is also known as a* **cap**.

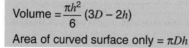

> Volume $=\frac{\pi h^2}{6}(3D - 2h)$ where D = diameter of sphere
>
> Area of curved surface only $= \pi Dh$ h = height of segment

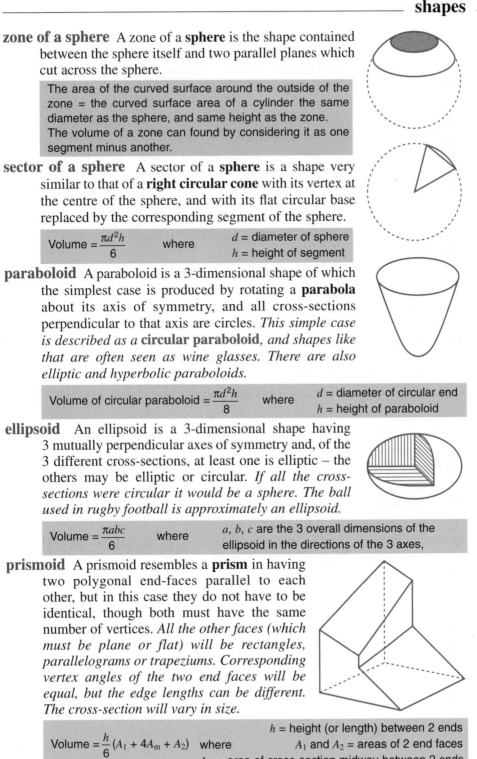

zone of a sphere A zone of a **sphere** is the shape contained between the sphere itself and two parallel planes which cut across the sphere.

> The area of the curved surface around the outside of the zone = the curved surface area of a cylinder the same diameter as the sphere, and same height as the zone.
> The volume of a zone can found by considering it as one segment minus another.

sector of a sphere A sector of a **sphere** is a shape very similar to that of a **right circular cone** with its vertex at the centre of the sphere, and with its flat circular base replaced by the corresponding segment of the sphere.

> $$\text{Volume} = \frac{\pi d^2 h}{6} \qquad \text{where} \qquad \begin{aligned} d &= \text{diameter of sphere} \\ h &= \text{height of segment} \end{aligned}$$

paraboloid A paraboloid is a 3-dimensional shape of which the simplest case is produced by rotating a **parabola** about its axis of symmetry, and all cross-sections perpendicular to that axis are circles. *This simple case is described as a* **circular paraboloid**, *and shapes like that are often seen as wine glasses. There are also elliptic and hyperbolic paraboloids.*

> $$\text{Volume of circular paraboloid} = \frac{\pi d^2 h}{8} \qquad \text{where} \qquad \begin{aligned} d &= \text{diameter of circular end} \\ h &= \text{height of paraboloid} \end{aligned}$$

ellipsoid An ellipsoid is a 3-dimensional shape having 3 mutually perpendicular axes of symmetry and, of the 3 different cross-sections, at least one is elliptic – the others may be elliptic or circular. *If all the cross-sections were circular it would be a sphere. The ball used in rugby football is approximately an ellipsoid.*

> $$\text{Volume} = \frac{\pi abc}{6} \qquad \text{where} \qquad \begin{aligned} a, b, c \text{ are the 3 overall dimensions of the} \\ \text{ellipsoid in the directions of the 3 axes,} \end{aligned}$$

prismoid A prismoid resembles a **prism** in having two polygonal end-faces parallel to each other, but in this case they do not have to be identical, though both must have the same number of vertices. *All the other faces (which must be plane or flat) will be rectangles, parallelograms or trapeziums. Corresponding vertex angles of the two end faces will be equal, but the edge lengths can be different. The cross-section will vary in size.*

> $$\text{Volume} = \frac{h}{6}(A_1 + 4A_m + A_2) \qquad \text{where} \qquad \begin{aligned} h &= \text{height (or length) between 2 ends} \\ A_1 \text{ and } A_2 &= \text{areas of 2 end faces} \\ A_m &= \text{area of cross-section midway between 2 ends} \end{aligned}$$

space and shapes

2-D or **two-dimensional space** A space is described as being 2-dimensional if, to give the position of any point in that space, 2 and only 2 measurements are necessary from a pair of non-parallel straight lines fixed in that space.

3-D or **three-dimensional space** A space is described as being 3-dimensional if, to give the position of any point in that space, 3 and only 3 measurements are necessary from 3 straight lines (no pair being parallel) fixed in that space.

shape A shape is made by a line or lines drawn on a surface, or by putting surfaces together. *It is usual in mathematics to require that the lines or surfaces are closed in such a way that an inside and an outside of the shape can be defined. When a shape is named it needs a context to determine whether it is the enclosed space that is being referred to or its defining outline.*
Example: The word 'circle' may refer either to the line defining it, or to the shape enclosed by that line.

plane shape A plane shape is a **shape** which is contained entirely within a simple flat (2-dimensional) surface or plane.

solid A solid is a **shape** formed in **3-dimensional space**. *The most common of these are the: cube, cuboid, cylinder, cone, pyramid, prism, and sphere.*

edge An edge of a **shape** is the line, or one of the lines, defining the outline of that shape. *In 3-dimensional shapes the edges are usually formed where the defining surfaces meet.*

face A face is a plane surface enclosed by an **edge** or edges.

side The word side is used to refer to the **edge** of a 2-dimensional shape.

vertex A vertex of a shape is a point at which 2 or more **edges** meet. *It is more commonly referred to as a 'corner'.*

diagonal A diagonal of a shape is a straight line which joins one **vertex** to another vertex and which is NOT an edge of that shape.

face diagonal A face diagonal of a 3-dimensional shape is a **diagonal** which lies entirely in one **face** of the shape.

space diagonal A space diagonal of a 3-dimensional shape is a **diagonal** which is NOT a **face diagonal**.

perimeter The perimeter of a **shape** is the total distance around the edges defining the outline of that shape.
Example: The perimeter of the shape on the right is found by adding together the lengths of the 4 edges marked a, b, c and d.

circumference ≡ **perimeter** of a circle or an ellipse.

rectilinear shape A rectilinear shape is a **shape** whose **edges** are all straight lines. *All polygons are rectilinear shapes.*

dihedral angle When two planes intersect in a common straight line, the dihedral angle between the planes is the one measured between two lines, one in each plane, drawn from the same point on that common line and perpendicular to it. *Dihedral angles are most often met with as the angle between two adjacent faces of a polyhedron.*

dimension   A dimension of a shape is one measurement taken between two specific points on the outline of the shape or, in some cases, inside the shape. *Usually several dimensions are needed to fix the size of a shape.*

length
breadth
width
height
depth
thickness

These are all commonly used labels to indicate positions for which the **dimensions** of a shape are given or needed. Conventional usage is:
• length for a single dimension
• length and breadth (or width) for 2 dimensions with length
being the greater, plus
• height, depth or thickness when a 3rd dimension is needed.
However, these labels are often used much more loosely.
Example: The three main dimensions of a cupboard are generally referred to as its width, depth and height.

area   The area of a surface is a measure of how much 2-dimensional space is covered by that surface. *It is usually measured in terms of how many squares of some unit size (square inches, square metres, etc.) would cover an equivalent amount of space. The surface may be flat or curved.*

volume   The volume of a 3-dimensional shape is a measure of how much space is contained within, or occupied by, that shape. *It is usually measured in terms of how many cubes of some unit size (cubic inches, cubic metres, etc.) would fill an equivalent amount of space.*

Reuleaux polygon   A Reuleaux polygon is similar in appearance to a **polygon** except that all its edges are curved in such a way that the distance between any pair of parallel lines touching the edges at opposite points is always the same. *It is also known as a* **curve of constant width**, *and is seen in the design of some coins. The diagram shows a* **Reuleaux triangle**. *The polygon used as the basis must have an odd number of vertices but does not have to be regular, though the final shape is then more difficult to draw. The vertices can also be rounded by drawing further suitable curves.*

ruled surface   A ruled surface is a **surface** on which, AT ANY POINT, at least one straight line can be drawn which lies entirely on that surface. *A plane or flat surface is the simplest case, but the curved surfaces of a cone and cylinder are also ruled surfaces, while a sphere is not. Some ruled surfaces can be surprisingly complex.*

geodesic line   A geodesic line between two points on the surface of a 3-dimensional shape is the line which also lies entirely on the surface of the shape and is of the shortest possible length. *On a sphere a geodesic line is an arc of the* **great circle** *which passes through the two given points, and this fact is of importance to navigators who wish to find the shortest distance between two places on the Earth's surface.*

concurrent   Two, or more, lines are said to be concurrent if they have one point in common, through which all the lines pass.

round angle   A round angle is an angle of 360 degrees or one complete revolution. *It is also known as a* **perigon**.

statistics involves the collection, display and analysis of information. *Usually the information is numerical in type or else is changed into a numerical form.*

data is the complete set of individual pieces of information which is being used in any of the processes connected with **statistics**.

raw data is the **data** as it was originally collected, before any processing at all has been done.

grouped data is **data** that has been put into groups according to some particular rules to make it easier to handle. *It is most often grouped according to size. Example: Collecting data about the heights of 100 people could result in 100 different measurements. It is easier to handle if the data is put into groups. Suitable groups might be: less than 100; 100–120; 120–140; 140–160; 160–180; 180–200; over 200 (all in cm). That is very simple but still needs decisions as to where a piece of data at the end of a group goes. For instance, to which group does a person of height 160 cm belong? For greater precision this is done by writing the class interval as falling between two inequalities such as '140 < height ≤ 160 cm'. In this case the first and last groups would be 'height ≤ 100 cm' and 'height > 200 cm'.*

class A class of data is one of the groups in a collection of **grouped data**.

class limits are the two values which define the two ends of a **class** and between which the data must lie.

class interval The class interval is the width of a **class** as measured by the difference between the **class limits**. *In any collection of grouped data the class intervals are very often all the same but do not have to be. Common exceptions are the classes at each end when arranged in size order. Example: Under 'grouped data' the example uses definite class intervals of 20 cm except for the first and last classes (or groups), which are open.*

discrete data is **data** which can only be of certain definite values. *Example: A survey of shoe sizes being worn by a group of people would be using discrete data, since there are only a limited amount of values for the sizes in which shoes are made and sold.*

continuous data is **data** which can take any value within certain restrictions. *The restrictions might be the class limits if the data is grouped; or those of the measuring device when the data is collected; or those of common sense. Example: A survey of lengths of people's feet involves continuous data. They can be of any length, but the usual measuring instruments do not record beyond 3 decimal places, and common sense dictates that there are certainly both upper and lower limits to what we might expect to find.*

frequency The frequency of some **data** is the number of times each piece of that data is found.

f is the symbol for **frequency**.

population A population is the complete set of objects (values or people) which is being studied by some statistical method.

distribution The distribution of a set of **data** is a graph or table showing the **frequency** of the data in each **class**, or of each type.
Example: This table shows the distribution of shoe sizes among 100 people:

Shoe size	2	3	4	5	6	7
Frequency	14	20	35	18	9	4

normal distribution A normal distribution is one in which the **frequency diagram** is symmetrical about a line through the **mean** value (which is also the **median** and the **mode**), and has a shape like that shown on the right. *The curve (known as a 'bell-curve') is derived from a particular formula. It really needs a lot of data to produce something that approximates to a curve, but the phrase is loosely used for frequency diagrams that seem to have that rough general shape.*

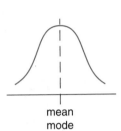

mean
mode

skewed distribution A skewed distribution is one in which the **frequency diagram**, whilst having a single mode, is not symmetrical about the mean.

positively skewed A distribution is said to be positively skewed if the **mode** lies to the LEFT of the **mean** (mode < mean) in the frequency diagram.

negatively skewed A distribution is said to be negatively skewed if the **mode** lies to the RIGHT of the **mean** (mode > mean) in the frequency diagram.

dispersion The dispersion of a set of **data** is a measure of the way in which the **distribution** is spread out. *There are various ways it can be measured but the one most often used is the one known as the **standard deviation**.*

spread ≡ **dispersion**

modal class The modal class of a set of **grouped data** is the **class** which has the greatest **frequency**.

bimodal A set of **grouped data** is said to be bimodal when the **distribution** (shown graphically) has two separate and distinct peaks.
Example: The drawing on the right shows a distribution which is bimodal.

two-way table A two-way table is a table of data which shows the combined effect of two separate happenings. *On the right is a two-way table which shows (in red) the combined total scores possible when 2 dice are rolled. From this table it can be seen that, with two dice, a score of 7 can happen in more ways than any other score, while 2 and 12 can happen in only one way.*

+		Second die				
	1	**2**	**3**	**4**	**5**	**6**
1	2	3	4	5	6	7
2	3	4	5	6	7	8
3	4	5	6	7	8	9
4	5	6	7	8	9	10
5	6	7	8	9	10	11
6	7	8	9	10	11	12

First die

statistics (general)

primary data is **data** which has been collected at first hand, usually by the individual(s) actually doing the statistics with that data.

secondary data is **data** which has been taken from an existing source.

qualitative data is **data** which is described by some quality that it possesses. *Two qualities often used are type and colour. The statistical measures of* **mean** *and* **median** *cannot be applied to this kind of data but the* **mode** *can. It is also known as* **categorical data**.
Examples: A survey of vehicles might separate them by type into cars, lorries, buses etc. A survey of cars might separate them by colour into red, blue, green etc.

quantitative or **quantitive data** is **data** which is described by means of a measurement of some kind. *The measurements most commonly used are those of length and weight. Most statistics is concerned with this kind of data. It is also known as* **numerical data**.

time series A time series gives several values of a measurement taken at different times. *It shows how the measurement varies with time, and is one of the most commonly seen types of graph in everyday use. It is often used to make predictions about what is going to happen.*

Example: The graph shows a possible set of values for a time series. No scales are marked. It could be showing the amount of radioactivity in a certain spot against

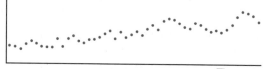

Time →

time measured in nanoseconds; or the amount of movement of a glacier (in metres) against time measured in centuries.

trend The trend of a **time series** is an indication of the general direction of its movement over the period of time for which the measurements were made. *The above graph shows an upward trend.*

variations All measurements taken for a **time series** show variations for many reasons, not all of which can be identified.

random variations in a **time series** are those **variations** which happen in an uncontrollable and unpredictable way throughout. *Such variations may be assumed to be present in any time series in addition to any known types.*

cyclical variations in a **time series** are those **variations** where the general shape of the graph has a tendency to repeat.

periodic variations in a **time series** are **cyclical variations** which happen at identifiable intervals throughout the time.

seasonal variations in a **time series** are **periodic variations** where the intervals can be matched to the seasons of the year.
Examples: The monthly figures for unemployment tend to be up in the winter and down in the summer. The sale of swimsuits goes the other way.

secular variations in a **time series** are **variations** which can only be observed after long periods of time. *'Long' is not defined, but is usually taken to be greater than 2 years, and may be several thousands of years.*

sample A sample is a set chosen from a **population** and used to represent that population in the statistical methods being applied. *This is necessary where it is not possible to collect all the data from a very large population.*
Example: In conducting an opinion poll to see how people would vote in an election, it is only possible to ask a sample of the population about their intentions and predict a result for the whole population from that.

random A result (of some experiment) is said to be random if, from all the results that could happen as a consequence of that experiment, each result has the same chance of happening and, no matter how many results have already been found, the next cannot be predicted.

random selection A random selection is any process by which objects (or numbers) are chosen in such a way that the appearance of each object is **random**. *The process might be rolling a fair die, reading numbers from a printed random number table – or drawing counters from a bag, provided that the counters are replaced and the bag is well shaken between each drawing.*
Computers and calculators usually have a routine for generating random numbers, although, strictly speaking, these should be called pseudo-random since the routine must depend upon a known algorithm.

random sample A random sample is a **sample** that has been chosen by a process of **random selection** from a **population**. *Note that it is not certain that such a sample is properly representative of the whole population.*

systematic sampling is the method used to produce a **sample** from a **population** which is arranged in some order. *The order might be actual or implied, as with birth dates.*
Example: From a list of 30 names, or a street of 30 houses, it is desired to choose one-third of them as a sample. Taking every 3rd name or house would be a systematic way of producing the sample of 10.

stratified sampling is done by dividing the population to be sampled into groups, or strata, according to some criterion, and taking appropriate samples (random or systematic) from each of those groups. These separate samples are then put together to make a stratified sample of that population. *Commonly used criteria are age, sex, social class, and occupation.*

quota sampling is done by deciding in advance how many of the population, in each of certain categories, are to be chosen. *The quotas are often set to represent how many of each category are known to be present in the total population.*
Example: A survey about attitudes among older persons might specify that, from the over-sixties, 100 men and 125 women are to be questioned.

sampling error is the difference between the **mean** of the **sample** and the mean of the **population** from which that sample was drawn. *It is important to have some idea of the probable size of this error in order to assess how much confidence can be placed in any conclusions made, based on the sample.*

statistics (graphical)

frequency diagram A frequency diagram is a graphical way of showing the amount of **data** found in each of the groups or types being counted.

bar chart A bar chart is a **frequency diagram** using rectangles of equal width whose heights or lengths are proportional to the frequency. *Usually adjacent rectangles or bars only touch each other if the data is continuous; for discrete data a space is left between the bars. The bars may be of any width and sometimes are no more than lines.*

block graph A block graph is a **bar chart** where, usually, the bars themselves are divided to mark off each individual piece of data.

histogram A histogram is a **frequency diagram** using rectangles whose widths are proportional to the **class interval** and whose areas are proportional to the frequency. *The class intervals may or may not be of equal width; if they are of equal width then the histogram is indistinguishable from a bar chart.*

pictogram A pictogram is a **frequency diagram** using a symbol to represent so many units of data. *The symbol usually relates to the data being shown.*

stem and leaf plot A stem and leaf plot is a **frequency diagram** which displays the actual data together with its frequency, by using a part of the value of each piece of data to fix the class or group (the stem), while the remainder of the value is actually listed (the leaves).

pie chart A pie chart is a circular **frequency diagram** using sectors whose angles at the centre are proportional to the frequency.

scattergram A scattergram shows how two sets of numerical **data** are related, by treating matching pairs of numbers as coordinates and plotting them as a single point, repeating this as necessary for each data-pair.

correlation is an assessment of how strongly two pieces of **data** appear to be connected to the extent that a change in one of them must produce a change in the other. *This assessment is usually made after a scattergram has been drawn. It can vary from being non-existent through weak to very strong.*

positive correlation is a **correlation** in which an INCREASE in the value of one piece of data tends to be matched by an INCREASE in the other.

negative correlation is a **correlation** in which an INCREASE in the value of one piece of data tends to be matched by a DECREASE in the other.

line of best fit The line of best fit is the **trend line** drawn on a **scattergram**. *The higher the correlation, the easier it is to draw this line.*

cumulative frequency is the total of all the **frequencies** of a set of **data** up to any particular piece or group of data.

cumulative frequency diagram or **polygon** A cumulative frequency diagram is a diagram on which all the various **cumulative frequencies** are plotted, each against the data value for which it has been calculated. *The points may be joined by straight lines (when it is usually called a polygon) or, if there are sufficient points to define it, by a curve.*

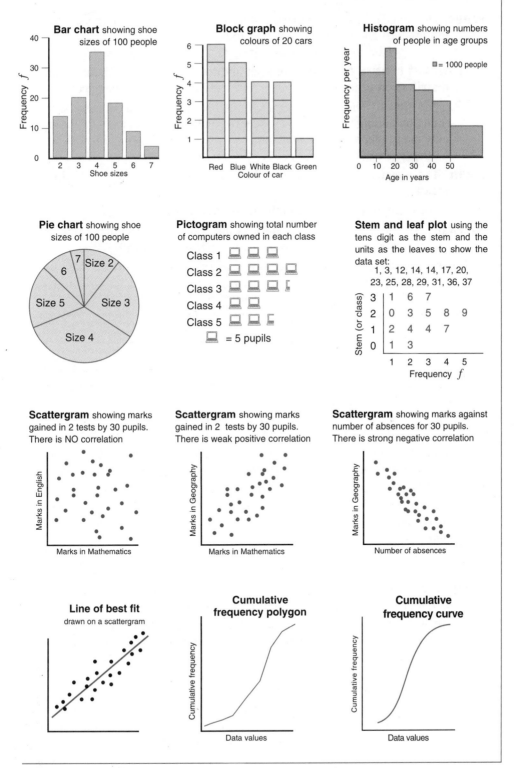

Bar chart showing shoe sizes of 100 people

Block graph showing colours of 20 cars

Histogram showing numbers of people in age groups
■ = 1000 people

Pie chart showing shoe sizes of 100 people

Pictogram showing total number of computers owned in each class

Class 1
Class 2
Class 3
Class 4
Class 5
= 5 pupils

Stem and leaf plot using the tens digit as the stem and the units as the leaves to show the data set:
1, 3, 12, 14, 14, 17, 20, 23, 25, 28, 29, 31, 36, 37

Scattergram showing marks gained in 2 tests by 30 pupils. There is NO correlation

Scattergram showing marks gained in 2 tests by 30 pupils. There is weak positive correlation

Scattergram showing marks against number of absences for 30 pupils. There is strong negative correlation

Line of best fit drawn on a scattergram

Cumulative frequency polygon

Cumulative frequency curve

range The range of a set of numerical **data** is the numerical difference between the smallest and the greatest values to be found in that data.
>*Example: For the data 9, 3, 3, 15, 11, the range is 15 – 3 = 12*

measures of central tendency of a set of **data** are any values about which the **distribution** of the data may be considered to be roughly balanced.

arithmetic mean The arithmetic mean of a set of **data** is the numerical value found by adding together all the separate values of the data and dividing by how many pieces of data there are. *It is a measure of central tendency.*
>*Example: For the data 9, 3, 3, 15, 11, the arithmetic mean is 41 ÷ 5 = 8.2*

mean The mean value of a set of **data** is usually taken to be the **arithmetic mean**.

$\bar{x}$ is the symbol for the **arithmetic mean** of a set of values of x.

average An average of a set of **data** is any **measure of central tendency**. *Usually it is taken to be the same as the **arithmetic mean**.*

weighted mean The weighted mean of a set of **data** is the **mean** value found after each piece of data has been multiplied by some factor which gives a measure of its importance or its **frequency** of happening.
>*Example: This table shows the distribution of shoe sizes among 100 people:*

Shoe size	2	3	4	5	6	7
No. of people	15	19	35	18	8	5

>*The mean shoe size could be given as (2+3+4+5+6+7) ÷ 6 = 4.5 but this does not allow for the fact that some sizes are a lot more common than others. The weighted mean size is*
>$[(2 \times 15) + (3 \times 19) + (4 \times 35) + (5 \times 18) + (6 \times 8) + (7 \times 5)] \div 100 = 4$

working mean A working mean is an assumed value for the **mean** of a set of **data**. *The use of a working mean allows other calculations to be done as the data is being entered and a correction made once the true mean is known.*

median The median value of a set of **data** is the numerical value of the piece of data in the middle of the set, AFTER ARRANGING THE SET IN SIZE ORDER. *If there is an even number of pieces of data, the **mean** of the middle two is taken as the median.*
>*Example: Data 9, 3, 3, 15, 11 has a median of 9 (middle of 3, 3, 9,11, 15) while 6, 2, 12, 4, 7, 18 has a median of 6.5 (mean of middle pair 6 and 7).*

mode The mode of a set of **data** is the piece of data found MOST often.
>*Example: Data 9, 3, 3, 15, 11 has a mode of 3 (as there are most 3's).*

percentile When a set of data is arranged in size order, the nth percentile is the value such that $n\%$ of the data must be less than or equal to that value; and n must be a whole number from 1 to 99. *Percentiles should only be used with large sets of data so that dividing it up into 100 equal parts (as the word 'percentiles' implies) seems realistic.*

>*Example: When the data (arranged in size order) is the set of measurements:*
>*3.7   4.5   7.3   8.3   8.4   9.6   10.1   10.8   11.6   12.4 cm, then the 30th percentile is 7.3 but that data set is unrealistic for percentiles.*

lower quartile The lower quartile of a set of data is the 25th **percentile**. *One quarter (25%) of all the data must have a value that is less than, or equal to, the value of the lower quartile.*
Example: For the data given under 'percentile' the lower quartile must be midway between the 20th and 30th percentile which is $(4.5 + 7.3) \div 2 = 5.9\,cm$.

upper quartile The upper quartile of a set of data is the 75th **percentile**. *Three-quarters (75%) of all the data must have a value that is less than, or equal to, the value of the lower quartile.*
Example: For the data given under 'percentile' the upper quartile is 11.2 cm.

interquartile range The inter-quartile range of a set of data is the difference in value between the **lower** and **upper quartiles** for that data. *It is one way of measuring the dispersion of the data.*
Example: The interquartile range for the above data is $11.2 - 5.9 = 5.3\,cm$.

semi-interquartile range = one-half of the **interquartile range**.
Example: The previous example has a semi-interquartile range given by
$$5.3 \div 2 = 2.65\,cm.$$

deviation The deviation of a value is the difference between that value and some other value. *The other value is usually the mean or median of the data.*

mean deviation The mean deviation of a set of **data** is the mean distance between the value of each piece of data and some fixed value. *The fixed value is usually the mean of all the data, but it can be the median. All the distances are considered to be positive.*

standard deviation The standard deviation of a set of **data** is a measure of its **dispersion** and is found by carrying out this calculation:

Find the difference in value between each piece of data and the mean of all the data.
Square all the differences *(this makes them all positive)*.
Add them together and divide by how many there are.
Take the square root.

Many calculators have a key that allows this procedure to be carried out automatically once the data has been entered.
*Example: Using the data for shoe sizes given under **weighted mean** and working from the mean of 4, the calculation is:*
$$[(^-2^2 \times 15) + (^-1^2 \times 19) + 0 + (1^2 \times 18) + (2^2 \times 8) + (3^2 \times 5)] \div 100 = 1.74$$
(The multiplications allow for the frequency of each piece of data.)
$$\text{So the standard deviation is } \sqrt{1.74} = 1.32 \text{ (to 3 sf)}$$

variance The variance of a set of **data** is a measure of its **dispersion**. Its value is given by the square of the **standard deviation**. *It can also be said that the standard deviation is the square root of the variance.*
Example: In the previous example the variance is $1.32^2 = 1.74$

σ (the Greek letter sigma) and s are symbols used for the **standard deviation**.
σ is used when it is for the whole population, s is used for a sample.

σ^2 and s^2 are symbols used for the **variance**.

box and whisker diagram A box and whisker diagram (also called a **boxplot**) is a drawing which displays seven measures relating to one set of data. *The diagram must be matched to a relevant number line, which is usually the 'horizontal' axis of the associated frequency diagram. The numerical values of the measures may, or may not, be written in as well. It is a useful pictorial summary when comparing sets of data.*
Example: This box and whisker diagram (drawn in red) summarises the data concerning the heights of a group of people.

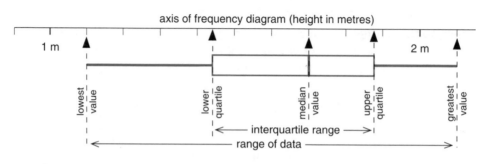

axis of frequency diagram (height in metres)

moving average A moving average is used to smooth out the fluctuations in value of a set of data which varies widely over time. It is made by re-evaluating the mean of the last few pieces of data whenever a new piece of data is added to the list. *It is used to give a clearer idea of the underlying trend.*
Example: A shop's recorded weekly sales of pencils might be:

8   10   6   11   4   9   9   6   5   10

which, by taking 3-weekly moving averages, gives this smoother looking set:

8   9   7   8   7.3   8   6.7   7

Notice that each value of the moving average is placed at the MIDDLE of the group of 3 pieces of data for which it has been calculated. The effect is most marked when the data is plotted onto a graph. This is done on the right where the black line is for the raw data, and the red line shows the moving average. There is a clear indication that sales are decreasing.

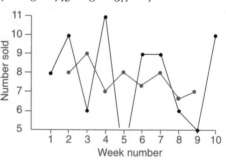

Spearman's rank order correlation coefficient is used when two sets of data have each been ranked in order (or the same set ranked in order by two different persons or methods) to give a measure of how well the two rankings agree. To calculate its value:

Find the difference in value between each corresponding pair of rankings.
Square all the differences *(this makes them all positive)*.
Add the squared values together and multiply by 6
Divide the previous result by $n(n^2 - 1)$, where n is the number of pairs.
Subtract the result from 1

Kendall's rank order correlation coefficient serves the same purpose as **Spearman's** but the coefficient is calculated by using the number and sizes of the changes needed to change one rank order into the other.

Pearson's product-moment correlation coefficient (r) is a measure of the linear **correlation** between two sets of data. *The two sets must be matched by means of a* **mapping** *showing a* **one-to-one correspondence***, meaning it can be displayed graphically as a* **scattergram***. Usually called* PMCC.

Identify each piece of data in one set as x-values; the others as y-values, so that for each x-value there is a corresponding y-value, and there are n pieces of data in each set.

To calculate r first carry out the instructions labelled **A** to **G**.

A. Multiply matching x- and y-values together; add them up; multiply by n.
B. Add up all x-values; add up all y-values; multiply the two results together.
C. Subtract result of **B** from **A**. (*This might be negative.*)
D. Square all x-values, add them up, multiply the total by n. Repeat for y-values.
E. Add together all x-values and square the total. Repeat for y-values.
F. Subtract the x-result in **E** from that in **D**. Repeat for y-results.
G. Multiply the two answers from **F** together and take the square root.

then r = Result from **C** ÷ Result from **G**

least squares The method of least squares is used to determine the position of a **line of best fit** for two sets of data plotted on a **scattergram**. *In this method (devised by* **Gauss***) the line is the one which ensures that the total sum of all the squares of the distances between the various pieces of data and that line is as small as possible.*

index numbers are used to compare the growth of some measurable quantity by studying the multiplier needed to make the new value from the old one. *The starting point is referred to as the base-value and all subsequent values are measured by comparison with that, so:* New value = base value × index. *This sort of index (which is a multiplier) should not be confused with* **index notation** *(where the index is an exponent). Index numbers are frequently used in economics; the best known is the* **RPI** *or* **Retail Prices Index**.

population pyramids are a particular example of how two **bar charts** can be used 'back-to-back' to display two sets of data against a common base-line for purposes of comparison. *The population pyramid on the right shows how many people there are in each of the given age-groups in the UK. The left-hand bar chart gives the figures for males, the right-hand one gives the figures for females. (Seven age groups have been left out to save space.) Such diagrams have important social implications.*

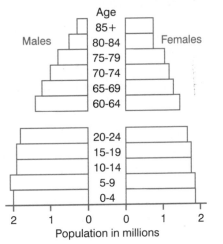

structures

structure The structure of a system first defines the objects to be used and then the way in which they are to be used. *The structure dealt with here is mainly that for arithmetic and the objects are numbers, but the general ideas and definitions are applicable to many other parts of mathematics.*

operation An operation is a rule (or body of rules) for processing one or more objects. *The most basic operations of arithmetic are those of addition, subtraction, multiplication and division, and the objects being processed are numbers. In mathematics generally the operations can be much more complex and the objects are usually algebraic in form.*

operator An operator is the symbol used to show which **operation** is to be done.
Examples: + − × ÷ are all operators.

set Before any **operation** can be used it is necessary to make clear on what **set** of objects it is to be used.
Examples: The set might be 'all whole numbers' or 'only positive numbers' or 'real numbers' or 'quadratic equations', and so on.

binary operation A binary operation is an **operation** which combines two objects to produce a third. *Addition, subtraction, multiplication and division are all binary operations, since they all require two numbers from which a third number is made.*

unary operation A unary operation is an **operation** requiring only one object to work on from which it produces another. $\sqrt{}$ *is a unary operation.*

closed An **operation** on a particular **set** is said to be closed if the operation always produces a result that is also in the set.
Examples: Addition on numbers is closed since number + number = number. Multiplication is also closed. Subtraction on positive numbers is not closed as it is possible to get a negative result which is not in the set. Division on integers is not closed because fractions are excluded.

commutative A commutative **operation** is one in which the order of combining the two objects does not matter.
Examples: Addition is commutative since 3 + 5 = 5 + 3. Subtraction is NOT commutative since 6 − 4 ≠ 4 − 6

associative A **binary operation** is said to be associative if, when it is being applied repetitively, the result does not depend on how the pairs are grouped (whether by working or by the insertion of brackets).
Examples: Addition is associative since 1 + 2 + 3 = (1 + 2) + 3 OR 1 + (2 + 3) = 6. Subtraction is not associative since 5 − 4 − 2 might be EITHER 5 − (4 − 2) = 3 OR (5 − 4) − 2 = ⁻1 producing two different results.

identity The identity for a **binary operation** on a **set** is an object in that set which, when combined (by the operation) with any second object from the set, produces a result which is equal to the second object.
*Examples: Addition of numbers has the identity 0 since 0 + 5 = 5
Multiplication of numbers has the identity 1 since 1 × 7 = 7
This is generally expressed as 0 + x = x + 0 = x and 1 × x = x × 1 = x*

left/right identity A left identity or a right identity is an **identity** which works only on the nominated (left or right) side of the other object.
Examples: Subtraction of numbers has a right identity of 0 since $x - 0 = x$ but no left identity since $0 - x \neq x$. Division has a right identity of 1

inverse The inverse of an object in a **set** under a **binary operation** is another object in the same set which, when combined with the original object, produces the **identity** as the result.
Example: For numbers under addition any number x has the inverse ^-x since $x + {}^-x = 0$ (the identity for addition).

group A group is a **set** under a **binary operation** for which ALL the following statements are true:

> The operation is **closed**.
> The operation is **associative**.
> The set has an **identity**.
> Every object in the set has an **inverse**.

Example: Integers under addition are a group but not under multiplication since inverses (= fractions) do not exist within the set.

commutative group A commutative group is a **group** with the extra property that the operation is **commutative**.

Abelian group ≡ **commutative group**.

distributive law The distributive law (if it is applicable) describes how two operators may be used together when linked in a particular way.
Example: The distributive law of arithmetic says that multiplication is distributed over addition as in $a \times (b + c) = a \times b + a \times c$

modulus The modulus of a particular system of arithmetic is an integer value which is used as a divisor throughout that system.

mod is an abbreviation for modulo which is used to identify the number being used as the **modulus**.
Example: When the modulus is 4 the system is working modulo 4 or mod 4

residue The residue of any number is the remainder after that number has been divided by a specified **modulus**.
Example: Using a modulus of 4 the residue of 7 is 3; or $7 \equiv 3 \pmod 4$.

congruent Two numbers are said to be congruent to each other if they both have the same **residue** when using the same **modulus**.
Example: 7 and 15 both have a residue of 3 after division by 4; so 7 is congruent to 15 (mod 4).

≡ is the symbol for 'is congruent to'. *Examples: $7 \equiv 3 \pmod 4$; $7 \equiv 15 \pmod 4$.*

modulo arithmetic is a system of arithmetic based on relating numbers to each other only by their **residues** for some given **modulus**.
Example: A multiplication table (mod 4) is shown. It is not a group as there are no inverses for 0 and 2: both $0 \times x = 1$ and $2 \times x = 1$ have no solution.

×	0	1	2	3
0	0	0	0	0
1	0	1	2	3
2	0	2	0	2
3	0	3	2	1

surveying is the science of measuring related to the Earth's surface and the presentation of those measurements in a suitable way. *It is used in making maps, measuring fields, and almost anything concerned with building on the Earth's surface – roads, reservoirs, houses etc.*

geodetic surveying is **surveying** done over large areas where the curvature of the Earth must be considered.

plane surveying is **surveying** done over small areas which can, without any serious errors, be treated as flat surfaces. *No limit on size can be given since it depends on what degree of error is acceptable, but a square of edge-length less than 15 km would be regarded as 'small' for most practical work.*

theodolite A theodolite is an instrument (usually mounted on a tripod) consisting of a sighting-tube (like a telescope) which can be moved in both the **horizontal** and **vertical** planes, with these movements (measured in degrees) being shown on two suitable scales. *It is the instrument people always associate with land-surveyors. Modern theodolites can be accurate to better than one-thousandth of a degree, often incorporate a means of measuring distance, and are capable of recording readings automatically.*

triangulation is the method of dividing up the surface which has to be measured into a set of triangles, so that each triangle is joined to at least two other triangles. *The usefulness of this method is that, from knowing only one edge-length, and most (but not all) of the angles involved, all the other edge-lengths can be calculated. Since angles can be measured more accurately and more easily than lengths, this method is used in all appropriate cases. In the diagram, the black outline marks the edges of the shape being surveyed and the red lines give the triangulation. If the length of the thick (red) line is known, together with the sizes of all the angles at each of the 5 marked (black) points, then the lengths of all the other lines can be calculated.*

traverse A traverse is a line which forms part of a **survey**. *Usually the length of the traverse is known. Sometimes the direction is also known.*

base line A base line is a line forming part of a **survey** (whose length may be known) from which other lines, and their measurements, are taken. *It is also known as a* **chaining line***.*

offset An offset is a line, and its measurement, which is taken off at right-angles to a **base line**. *The diagram on the right shows a (thick red) base line with (thin red) offsets and their lengths, used to fix the position of the black line which represents the boundary of the space being surveyed. The figures along the base line give the distance from one end. The trapeziums formed in this method are a help in calculating the area. The economical way in which all the measurements can be recorded is shown in the table to the left of the drawing.*

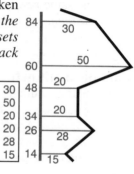

84	30	48
60	50	
48	20	34
34	20	26
26	28	
14	15	14

compass and traverse is a method of fixing the shape and size of the boundary to a shape by considering the boundary as a series of **traverses** and measuring the length and direction (by taking a compass **bearing**) of each one. *These measurements then allow an accurate drawing to be made. Bearings are taken at both ends of each traverse as a check.*

radial survey A radial survey is done by: establishing a fixed point inside the shape to be surveyed; running **traverses** from that fixed point to selected points on the edges of the shape; and then measuring the length of each traverse and the angles between them. *This method is easy to use and very suitable for fields and similar shapes which are reasonably flat.*

alidade An alidade is an instrument like a ruler with sights fixed at each end and is used to copy a direction on to a piece of paper. *In use the observer rests it on the paper, lines up the sights to give the direction of some distant object, and then draws a line on the paper.*

plane table A plane table survey is a type of **radial survey** which does not require the angles to be measured. Paper is fixed to a horizontal flat surface (like a drawing-board) placed above a fixed point on the ground and then, using an **alidade**, the direction of identifiable points on the boundary are drawn directly onto the paper. The table is then moved a measured distance to a second fixed point and set up facing the same compass direction as before. The position of the second fixed point is plotted on the paper (to scale) and another set of direction lines are sighted and drawn through that. The intersections of the two sets of directions fixes the relative positions of all the identifiable points. *It is a cheap and easy method but needs care.*

datum In measuring heights it is necessary to have a fixed horizontal line (or plane) from which all measurements (up or down) can be taken. This fixed line is known as the datum. In order that heights can be compared over an entire country there has to be one overall recognised datum and, for Great Britain, this is the Mean Sea Level measured at Newlyn in Cornwall. *It was established by taking hourly readings of the sea level over a 6-year period.*

benchmarks Though it is necessary to have only one national **datum** in a country it is not very convenient for a surveyor who is working some miles away from it. So, benchmarks are set up all over the country which give the height of that point above (rarely below) the national datum. *Benchmarks look like that in the drawing on the right, and are cut into the walls of bridges, churches etc.*

tacheometry is the name of the method by which the distance from a **theodolite** to some distant point is calculated. The measurement of the angle **subtended** at the theodolite (T) by an object of known length placed at the distant point allows T⊕ the distance to that point to be calculated.

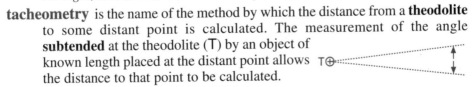

symbols

A symbol is a letter or sign used to represent instructions, or a number, in a more concise form. Sometimes a symbol replaces a group of words and can be read directly as it occurs, in these cases the words are shown below in 'quotation marks'.

$+$	'add' or 'plus' or 'positive'	
$-$	'minus' or 'subtract' or 'negative'	
$\sim$	find the absolute difference of	*Example:* $2 - 5 = {}^-3$ *but* $2 \sim 5 = 3$
$\times$	'times' or 'multiplied by'	
$*$	'times' or 'multiplied by'	*Usually used on a computer*
$\div$	'divided by' or 'shared by'	*Example:* $6 \div 3$ *or* $6/3$ *both mean '6 divided by 3'*
$/$	'divided by' or 'shared by'	*or '6 shared by 3' or 'How many 3's in 6?'*
$\pm$	'add or subtract' 'plus or minus' 'positive or negative'	*Example: When* $x = 3$ *then* $x \pm 2$ *gives the two answers 5 and 1* ± 6 *means the two numbers* $+6$ *and* ${}^-6$
$=$	'equals' or 'is equal to'	*Example:* $x + 3 = 7$
$\neq$	'does not equal' or 'is not equal to'	*Example:* $4 + 3 \neq 6$
$\approx$	'is approximately equal to'	*Example:* $\pi \approx 3.14$
$\equiv$	'is equivalent to' or 'has the same value as'	*Example:* £5 $\equiv$ 500p
$\equiv$	'is identically equal to'	*Example:* $(x + y)^2 \equiv x^2 + 2xy + y^2$
$<$	'is less than'	*Example:* $x < 5$ *means x can take any value which is less than 5 but cannot equal 5*
$\leqslant$	'is less than or equal to'	*Example:* $x \leqslant 7$ *means x can take any value which is less than 7 or may equal 7*
$>$	'is greater than'	*Example:* $x > 3$ *means x can take any value which is greater than 3 but cannot equal 3*
$\geqslant$	'is greater than or equal to'	*Example:* $x \geqslant 6$ *means x can take any value which is greater than 6 or may equal 6*
$\propto$	'varies as' or 'is proportional to'	*Example:* $y \propto x$ *means* $y = kx$ *where k has a constant value.*
$.$	decimal point or fraction point	*It is placed on the line and used to separate the whole number part from the fractional part.*
$,$	decimal marker	*Example:* 3,48 *is equivalent to* 3.48 *The comma is standard in the SI or metric system.*
$\%$	'per cent' or 'out of a hundred'	
$\%o$	'per mil' or 'out of a thousand'	
$[x]$	the largest whole number which is not greater than x	*Example: [3.5] is 3, but [−4.2] is* ${}^-5$
$\lvert x \rvert$	the value of x with no sign attached	*Example:* $\lvert {}^-8.7 \rvert$ *is 8.7*
x^2	'x squared' or x multiplied by itself	*Example:* $4^2 = 4 \times 4 = 16$
x^3	'x cubed'	*Example:* $2.5^3 = 2.5 \times 2.5 \times 2.5 = 15.625$

$\sqrt{x}$ 'the square root of x'   *Example:* $\sqrt{9} = 3$ *since* $3 \times 3 = 9$

$\sqrt[3]{x}$ 'the cube root of x'   *Example:* $\sqrt[3]{8} = 2$ *since* $2 \times 2 \times 2 = 8$

$\angle$ 'angle'   *Example:* $\angle ABC$ *or* $\angle B$

$\parallel$ 'is parallel to'   *Example:* $AB \parallel CD$

$\perp$ 'is perpendicular to'   *Example:* $AB \perp CD$

⌐   these lines are at right angles to each other

                               Examples:

   these lines are parallel to each other

   these lines are equal to each other in length

$\circ$ 'degree'   *of angle or temperature*

$'$ 'minute'   *one-sixtieth of a degree or hour*

$''$ 'second'   *one-sixtieth of a minute (of angle or time)*

$n\,!$ 'factorial n'   *Multiply together all the integers from 1 to n*

$\{\}$ used to enclose a listed set, or definition   *Example:* $\{A, B, C, D, E, F\}$

$\in$ 'is a member of'   *Example:* $D \in \{A, B, C, D, E, F\}$

$\notin$ 'is not a member of'   *Example:* $X \notin \{A, B, C, D, E, F\}$

$\subset$ 'is a sub-set of'   *Example:* $\{A, D, E\} \subset \{A, B, C, D, E, F\}$

$\supset$ 'includes'   *Example:* $\{A, B, C, D, E, F\} \supset \{A, D, E\}$

$\cup$ union of two sets   *Example:* $\{A, D\} \cup \{B, C, D, E\}$

                                    $\equiv \{A, B, C, D, E\}$

$\cap$ intersection of two sets   *Example:* $\{A, C, D\} \cap \{B, C, E\} \equiv \{C\}$

$\emptyset$ null or empty set   *Example:* $\{A, D\} \cap \{B, C, E\} \equiv \{\} \equiv \emptyset$

σ standard deviation

$\Rightarrow$ 'implies'

$\Leftarrow$ 'is implied by'

$\Leftrightarrow$ 'implies and is implied by'   *also written as* iff

$\therefore$ 'therefore'   *Example:* $AB \perp BC \therefore \angle ABC = 90°$

∞ 'infinity'

e ≈ 2.71828

$f(x)$ 'a function of x'   *Example:* f(x) *is* $x^2 + 3x - 4$

π pi ≈ 3.14159

$\&$ hexadecimal number follows   *Example:* $\&10A3C$ ($\equiv 68156$ *decimal*)

$(\,)$ round brackets   **Brackets** are used (in pairs) to enclose an expression that is to be

$[\,]$ square brackets   treated as a complete quantity and evaluated before the rest of the

$\{\,\}$ curly brackets   expression. Brackets can be nested within each other and the use
                   of different types allows matching pairs to be seen more readily.
                   Example: $4\{5x[2x(x+5) - 7]+8\} - 17$ *is* $40x^3 + 200x^2 - 140x + 15$

Greek alphabet The Greek alphabet is a rich source of symbols used in both mathematics and science, to the extent that nearly every one of them (both capitals and lower case) is used in some way or other. Some of them appear more than once to represent different things. Below is the full alphabet, and the names of the various symbols, with a note of where some of them are used. More detail can be found through the cross references given in **bold**.

A α alpha

α, β, γ are often used to identify angles *Example:*

$$\frac{\alpha/\beta}{\beta/\alpha}\quad\frac{\beta/\alpha}{\alpha/\beta}\quad\alpha + \beta = 180°$$

B β beta

Γ γ gamma

Δ δ delta

Δ is sometimes used to represent the area of a **triangle**.

δ is used to show that a small measure is to be taken. *Example: δy would mean 'a small amount of y'.*

E ε epsilon

Z ζ zeta

H η eta

Θ θ theta

θ is used to indicate a general angle as in **polar coordinates**.

I ι iota

K κ kappa

Λ λ lambda

λ is used to represent a **scalar** in vector work.

M μ mu

μ is used in the **SI** system of units to represent the prefix *micro*. *Example: μm is a micrometre, or one-millionth of a metre.*

μ is sometimes used to represent the **arithmetic mean**.

N ν nu

Ξ ξ xi

ξ is sometimes used as the symbol for a **universal set**.

O o omicron

Π π pi

Π is used to show that a **continued product** is needed.

π is used to represent the **irrational number** 3.14159...

$\pi(n)$ means the number of primes equal to, or less than, n. *Example: $\pi(13)$ is 6. The six primes are 2, 3, 5, 7, 11, 13*

n	$\pi(n)$	n	$\pi(n)$	n	$\pi(n)$
10	4	10 000	1229	10 million	664 579
100	25	100 000	9593	100 million	5 761 455
1000	168	1 million	78 499	1 billion	50 847 534

P ρ rho

Σ σ sigma

Σ is used to show that the sum of a **series** is to be found.

σ is used to represent the **standard deviation** of a population.

T τ tau

τ is used to represent the value of the **golden ratio** 1.6180...

Y υ upsilon

Φ ϕ phi

Φ is sometimes used (incorrectly) to represent an **empty set**.

X χ chi

Ψ ψ psi

Ω ω omega

+ − One of the earliest signs used to show that two numbers had to be added was the Ancient Egyptian hieroglyph $\wedge$ which represented a pair of legs walking forward (their writing and reading was done from right to left). So $\cap||\wedge|||$ would have been read as 3 + 12. Unsurprisingly, their minus sign was a pair of legs walking in the opposite direction. Up until the 1500s a variety of signs were used, but very often the instruction was written in full. Italian mathematicians of the 1400s used *p* and *m* (for *plus* and *minus*) which was a shortened form of their (Italian) words. The first + and − signs appeared in 1481 in a German manuscript on algebra. For quite some time their use appears to have been restricted only to algebra and it took nearly 100 years before they came into more general use in arithmetic.

× As with the + and − signs, there was a wide diversity of symbols used for multiplication. Many writers used lines to show which numbers were to be connected in some way and, inevitably, some of these lines crossed, so while the × appeared in many places it did not always mean that multiplication had to be done. It was the English mathematician William Oughtred who (in 1631) gave it the particular meaning it has today. Some objected it was too much like an *x*, but, nevertheless, it was slowly adopted over the next century.

÷ This was once commonly used as a minus sign. Then a Swiss mathematician, Johann Rahn, used it to show division in 1659 and it was adopted in England and the USA, but Continental Europe continues to use the colon : (as introduced by **Leibniz**) to show division.

= The sign for equality we use today was introduced by **Recorde** in his book *The Whetstone of Witte (1557)* where he justified it by explaining *'a paire of paralleles, lines of one lengthe, thus:* ══, *bicause noe.2.thynges, can be moare equalle.'* It was not immediately adopted by everyone. As with much mathematical notation of that period, everyone had his (very rarely her) own system, but by about 1700 the = sign was in universal use.

√ Early writers (pre-1500) used a dot to show that a root was to be taken. (The dot as a decimal point came later). It can only be conjectured that a 'tail' was added to the dot to make it more visible, because that is what some writers started to do. The present form of the sign was introduced in 1525 and was gradually taken into use over the next 100 years or so.

. **decimal fractions** were known about from very early times, though they were not used a lot even where they would have simplified the calculation. Their regular use was helped when Simon Stevin published (in 1585) a very clear description of them. However, he used a clumsy notation and it was **Napier** (in 1616) who introduced a decimal separator (comma or point) which made decimal fractions much easier to use.

! The **factorial** sign, like most mathematical symbols, has been represented in many different ways. Some of the simpler ones were n^* $\Pi(n)$ $[n]$ $\underline{|n}$ all representing what we now write as $n!$ The use of the exclamation mark was introduced in Germany in the early 1800s, but it was not until the middle of the 1900s that it could be said to be in universal use.

symmetry The word 'symmetry' applied to any object (or situation) means that parts of the object correspond to (or match) other parts in some way.

symmetry of shape The symmetry of a shape describes how, under certain rules of movement, the shape fits exactly on to itself. There are three types of shape symmetry: line, rotational and plane.

line symmetry is the **symmetry** of a plane shape *(= flat or 2-D shape)* which can be folded along a line so that one half of the shape fits exactly on the other half. *A shape can have several lines of symmetry.*
Examples with lines of symmetry shown in red:

rotational symmetry is the **symmetry** of a shape which may be turned and fitted on to itself somewhere other than in its original position.
Examples with centres, about which the shape is turned, shown in red:

centre of symmetry The centre of symmetry is the point about which a shape having **rotational symmetry** is turned.

point symmetry ≡ **rotational symmetry**

order of rotational symmetry The order of rotational symmetry of a shape counts the number of times that a shape can be turned to fit on to itself until it comes back to its original position. *Every shape has an order of rotational symmetry of at least 1, but this is usually ignored. In these examples the letters are used only to show the positions of the shape as it turns.*

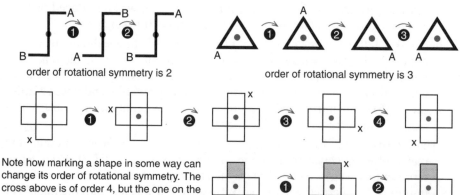

order of rotational symmetry is 2   order of rotational symmetry is 3

Note how marking a shape in some way can change its order of rotational symmetry. The cross above is of order 4, but the one on the right (with colour added) is of order 2.

plane symmetry is the **symmetry** of a 3-dimensional shape in which a plane (= *flat*) mirror could be placed so that the reflection looked exactly the same as the part of the shape being covered up by the mirror. *A shape can have several planes of symmetry.*
Examples, with planes of symmetry shown in red, are:

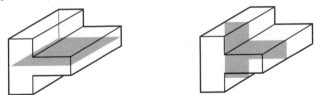

bilateral symmetry⎫ ⎧ all of these are the same as either,
mirror symmetry ⎬ ⎨ **line symmetry** of a 2-dimensional shape
reflective symmetry⎭ ⎩ or **plane symmetry** of a 3-dimensional shape.

asymmetric A shape having NO **symmetry** at all is described as asymmetric.
Examples of shapes that are asymmetric are:

axis of symmetry An axis of symmetry of a shape is a line about which the shape can be rotated, by an amount which is less than a whole turn, so that the total effect is to leave the shape unchanged. *A shape can have more than one axis of symmetry. In the case of line symmetry the line is the axis; for rotational symmetry the axis is a line passing through the centre of symmetry and perpendicular to the plane of the shape.*

axis of rotation An axis of rotation is a line about which a shape or another line is turned.

symmetric expression A symmetric expression is an algebraic expression using two or more variables whose value is unchanged if any two of the variables are interchanged.
Examples: $x + y$ $\qquad x^2 + y^2$ $\qquad x^2 + y^2 + z^2 + xyz$

symmetric equation A symmetric equation is an **equation** using a **symmetric expression**.
Examples: $x + y = 6$ $\qquad x^2 + y^2 = 10$ $\quad x^2 + y^2 + z^2 + xyz = 104$

symmetric function A symmetric function is a **function** using a **symmetric expression**. *Such functions when drawn as graphs will show some form of symmetry.*
Examples: $f(x) \equiv x + y$ $\quad f(x) \equiv x^2 + y^2$ $\quad f(x) \equiv x^2 + y^2 + z^2 + xyz$

symmetric relation A symmetric relation is a relationship which is true (or false) whichever way it is read. *The most commonly encountered symmetric relation is the equals sign since, if $x = y$, then $y = x$. Another is 'is parallel to' since, if AB is parallel to CD, then CD is parallel to AB.*

technical drawing

scale When making a drawing of an object which is meant to be in proportion to the size of the object, and from which measurements can be taken, a scale is used to fix the ratio between the actual measurements on the object and those in the drawing.

Scale is usually stated in one of two (or both) forms: either as a ratio or as a statement of how one measurement is related to the other.

1:1 means the drawing is the same size as the object.

1:10 means 1 cm on the drawing represents 10 cm on the object.

1:100 means 1 cm on the drawing represents 100 cm or 1 metre.

1:25 000 means 1 cm represents 25 000 cm or 250 metres.

The last one could also be stated as 4 cm represents 1 km. This is a scale commonly used on maps where it is also given as being equivalent to 2.5 inches to the mile; it is very close to this with an eror of only 1.4%

There might also be a **scaled ruler** *included on the drawing so that actual sizes can be read off directly with no need for any calculation.*

isometric drawing An isometric drawing tries to show a 3-dimensional view of an object in a 2-dimensional drawing. For this, the faces to be seen are plotted within a framework of 3-dimensional coordinates, and all lengths are measured (to **scale**) along, or parallel to, one of these three axes. On the paper the vertical axis is drawn vertically, and the two horizontal axes are drawn on either side of this vertical line and at an angle of 60° to it. *An isometric view of a house is shown.*

plan A plan drawing of an object is the 2-dimensional horizontal view that is seen when the object is looked at from a position above the object and looking straight down. *The drawing on the right is the plan that would be made for the house shown above.Usually plans are drawn to* **scale** *so that measurements can be taken from them.*

elevation An elevation drawing of an object is the 2-dimensional vertical view seen when the object is looked at from a position to one side of the object and looking straight at it. *Usually elevations are drawn to* **scale** *so that measurements can be taken from them, and are further identified as 'front', 'side', or 'end' elevations. On the right are two elevation drawings for the house drawn above.*

End elevation   Front elevation

golden ratio The golden ratio is used to divide an object into 2 parts so that the ratio of the larger to the smaller part is the same as the ratio of the whole object to the larger part. *Its value is the irrational number* $(\sqrt{5} + 1) \div 2$ *which is 1.6180 (to 4 decimal places). This ratio often appears in architectural design; it is believed such a division, where appropriate, is more pleasing to look at. The simplest example is the* **golden rectangle**.

bisect To bisect an object, usually a line, a shape or an angle, is to cut, or divide, it into two parts which are equal in size and shape.

perpendicular bisector A perpendicular bisector is a line which **bisects** another line and is at right angles to it. *The diagram shows the perpendicular bisector of the line* AB. *It cuts* AB *at* O *so* AO *must be equal to* OB *in length. Any point on the perpendicular bisector is at the same distance from either end of the line it bisects.*

mediator A mediator is most often considered to be a **perpendicular bisector** of a line but is, more generally, any **axis of symmetry**.

drop a perpendicular To drop a perpendicular from a (given) point on to a (given) straight line, is to draw a line from, or through, that point so that it forms a right angle with the line. *The diagram shows the perpendicular dropped from point* P *onto the line* XY. *It is assumed that the given point is not on the given line. In ordinary* **(Euclidean)** *geometry, from any one given point to any one line only one perpendicular exists. The discovery (in the 1800s) that this was not true for all geometries took mathematicians by surprise.*

foot of a perpendicular The foot of a perpendicular is that point at which it meets the line with which it forms a right angle. *In the above diagram the foot of the perpendicular from* P *to* XY *is the point* F.

apothem An apothem is a line drawn from the centre of a regular polygon to an edge, and perpendicular to that edge. *It is the perpendicular bisector of that edge, and also the radius of the inscribed circle to that polygon. The diagram shows one apothem (out of the 5 possible) of a regular pentagon.*

produce The instruction to 'produce the line …' means to extend it, or make it longer, in order to match certain conditions.
Example: 'Produce the line AB so as to meet the line XY'

circumscribe A second shape is said to circumscribe a first shape if the second completely encloses the first, generally touching it at several points, but not cutting it. *It is usual to require the circumscribing shape to be the smallest possible under the conditions given. The most common examples are the circumcircle to a polygon and the circum-sphere to a polyhedron. The drawing on the right shows a regular hexagon circumscribing an irregular quadrilateral.*

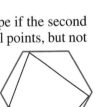

inscribe A second shape is said to be inscribed in a first shape if the second is completely inside the first, generally touching it at several points, but not cutting it. *It is usual to require the inscribed shape to be the largest possible under the conditions given. The most common examples are the incircle to a polygon and the in-sphere to a polyhedron. The drawing above shows an irregular quadrilateral inscribed in a regular hexagon.*

techniques

algorithm An algorithm is a step-by-step procedure that produces an answer to a particular problem. *Many algorithms are prepared ones which do the standard operations needed in an efficient or the most memorable way: multiplication, division, adding fractions etc. Other algorithms are devised as they are needed, particularly for computers.*

unitary method The unitary method can be used when the relationship between two quantities is fixed; the value of a third quantity is given; and the value of a fourth quantity has to be found having the same relationship to the third as the first pair have to each other. *The method is based upon making one of the first pair of quantities of unit value.*
Example: A car goes 480 km on 30 litres of petrol. (This is the first pair of quantities.) How much petrol will be needed for a journey of 1000 km? The working on the right shows how dividing the first pair by 480 produces a unit quantity and then multiplication is used to get the value of the fourth quantity.

```
480 km  : 30 litres
÷ 480     ÷ 480
1 km    : 0.0625 litres
× 1000    × 1000
1000 km : 62.5 litres
```

rule of three The rule of three applies to the same type of problems as described in **unitary method** but solves them by writing the given quantities down in a prescribed order, multiplying together the third and the second and dividing by the first.
Example: Using the same problem as given for the unitary method, the rule of three solution would be written out as:

$$480 : 30 :: 1000 : ? \qquad \text{(1st : 2nd :: 3rd : 4th)}$$
$$1000 \times 30 \div 480 = 62.5 \qquad \text{(3rd} \times \text{2nd} \div \text{1st} = \text{4th)}$$

laws of indices are those rules which control the operations of combining numbers written in index notation. These laws can only be applied to numbers in index form which have the same base.

With a base of b and index values of m, n then

$$b^m \times b^n = b^{m+n} \qquad b^m \div b^n = b^{m-n} \qquad (b^m)^n = b^{mn}$$
$$b^0 = 1 \qquad\qquad b^{-n} = \frac{1}{b^n}$$

Examples: $2^3 \times 2^5 = 2^8 \qquad 2^3 \div 2^7 = 2^{-4} \qquad (2^3)^5 = 2^{15} \qquad 2^{-3} = \frac{1}{2^3} = \frac{1}{8}$

iteration An iteration is a procedure which is repeated many times so that, from an estimated solution to a particular problem, each repeat produces a better approximation to the solution. *These solutions are usually numbered by means of a subscript, as in* $x_1 \quad x_2 \quad x_3 \quad x_4 \quad \dots \quad x_n \quad x_{n+1} \qquad$ *etc.*
Example: An iterative formula to find the cube root of a number (N) is:

$$x_{n+1} = \sqrt{\sqrt{N \times x_n}}$$

Given $N = 4$ *and starting with* $x_1 = 1$
then $x_2 = 1.4142...$ $\qquad x_3 = 1.5422... \quad x_4 = 1.5759...$
and so on to $x_{13} = 1.587401...$ *which is accurate to 6 decimal places.*

cross-multiplication is a method of simplifying an equation involving fractions based on the fact that $\frac{a}{b} = \frac{c}{d} \Leftrightarrow ad = bc$.

Example: $\frac{2}{5} = \frac{4}{x} \Leftrightarrow 2x = 5 \times 4$ *so* $x = 10$

$$\frac{a}{b} \diagdown\!\!\!\!\diagup \frac{c}{d}$$

area under a curve The area under a curve is the area enclosed between a curve drawn on a coordinate grid, two limiting **ordinates** and the *x*-axis. *Three methods of finding an approximate value for this area (apart from counting squares) are given.*

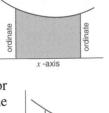

mid-ordinate rule The mid-ordinate rule is a method for finding the approximate area under a curve by dividing the space into strips; finding the areas of the separate strips by multiplying the length of the middle ordinate of each strip by the width of that strip; and finally adding the areas of the strips together. *It is a practical method that can be used when the equation of the curve is not known.*

trapezium rule The trapezium rule is a method for finding the approximate area under a curve which is similar to the mid-ordinate rule but treats each strip as a trapezium. *It can be very accurate if the strips are made as narrow as is practicable. By using strips of equal widths, the whole procedure can be reduced to a formula.*

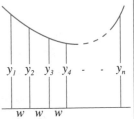

If the width of each strip is w and the lengths of the ordinates are y_1 y_2 y_3 .... y_n then the area is given by:
$$w \left[\tfrac{1}{2} (y_1 + y_n) + y_2 + y_3 + y_4 + \ldots + y_{n-1} \right]$$

trapezoidal rule ≡ **trapezium rule**

Simpson's rule is a method for finding the approximate area under a curve by dividing the space between the limiting ordinates into an even number of equal width strips. *It is usually more accurate than the previous methods.* With ordinates numbered from 1 to n (n being odd) the area is given by: width × [4 × sum of even ordinates + 2 × sum of odd ordinates − (first + last)] ÷ 3 *With any of these methods, if the equation of the curve is known, the lengths of the ordinates (the y-values) can be calculated, rather than measured, for the greatest accuracy.*

linear programming is a method used to find the 'best' solution to problems which can be expressed in terms of linear equations or inequalities. *Solutions are usually found by drawing graphs of inequalities and looking for optimum values that satisfy the required conditions. This method is widely used in business and industrial contexts and the problems often relate to obtaining maximum profits for given costs and production levels.*

solving quadratics If a quadratic equation is put in the form $ax^2 + bx + c = 0$, then it can be solved, and its roots found, by using the formula on the right.
$$x = \frac{-b \pm \sqrt{b^2 - 4ac}}{2a}$$

difference of two squares Any algebraic expression of the form $a^2 - b^2$ can be factorised into $(a + b)(a - b)$.
Example: $x^2 - 9$ is $x^2 - 3^2$ which is $(x + 3)(x - 3)$.

temperature The temperature of something is a measure of how hot it is according to a value given on a known scale. *Temperature is measured by means of a thermometer or similar instrument. When heat moves between objects it always moves from an object at a higher temperature to an object at a lower one. There are three main scales used for measuring temperature.*

Celsius scale The Celsius scale sets the freezing point of water at zero degrees [≡ 0°C] and the boiling point at 100 degrees [≡ 100°C].
It was devised by Anders Celsius (1701–1744), a Swedish astronomer.

Fahrenheit scale The Fahrenheit scale sets the freezing point of water at 32 degrees [≡ 32°F] and the boiling point at 212 degrees [≡ 212°F]. *Note that Fahrenheit degrees are smaller than Celsius degrees.*
It was devised by G D Fahrenheit (1686–1736), a German physicist.

To change temperatures between the Celsius and Fahrenheit scales use either the dual conversion scale on the far right or the conversion graph opposite. For more accuracy use one of these flow diagrams:

To change °F to °C use:

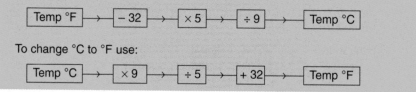

To change °C to °F use:

Kelvin scale The Kelvin scale is based on laws of physics in which there is an absolute zero, and all temperatures are measured from that point in units called kelvins, where 1 kelvin is the same size as 1 degree on the **Celsius scale**. *Note that when temperatures are given in K the ° sign is not used. Since the starting point is absolute zero, there can be no negative temperatures on this scale. In fact, temperatures of less than 0 K (zero kelvin) or its equivalent on any other scale, (⁻273.15°C or ⁻459.67°F) cannot exist. This scale was devised by William Thomson, 1st Lord Kelvin (1824–1907) a Scottish physicist and mathematician.*

To change temperatures between the Celsius and Kelvin scales use:
Temperature in °C = Temperature in K – 273.15
Temperature in K = Temperature in °C + 273.15

centigrade The centigrade scale was the name originally given to the **Celsius scale**. *It was officially changed (in 1948) because it could be confused with a system of angle measurement which used grades and centigrades.*

There were two other scales of note which were used for a while but are now obsolete. One was the Réaumur scale which was similar to the Celsius scale, with the same zero point but with only 80 degrees to the boiling point of water.
R A F de Réaumur (1683–1757) was a French entomologist.
The other was the Rankine scale which was similar to the Kelvin scale but based on the size of the Fahrenheit degree.
W J M Rankine (1820–1872) was a Scottish engineer.

conversion graph   A conversion graph is used to show the corresponding values between two quantities, which have a fixed relationship between them, by means of a line drawn on a squared grid. *In making a conversion graph the two quantities which are related are marked on a squared grid, using any suitable scales and putting one along each axis. A line is then plotted and drawn which shows the relationship between the two quantities so that intermediate values can be read off. The graph below can be used to change temperatures between the Celsius and Fahrenheit scales. The line of red dashes show that the readings of 35°C and 95°F are measuring the same temperature.*

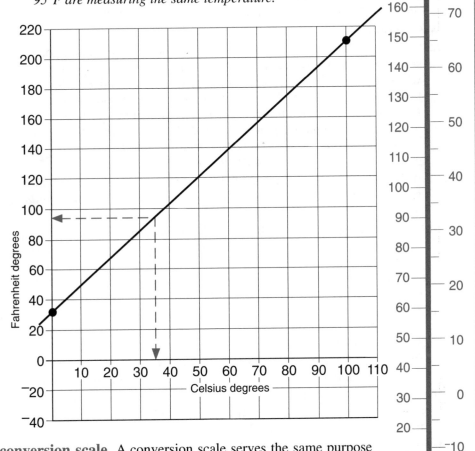

conversion scale   A conversion scale serves the same purpose as a **conversion graph** but the two scales are now matched directly side by side. *A conversion scale for changing between °F and °C is given on the right. It is more difficult to draw since both scales have to be matched along their entire lengths, but it is easier to use.*

137

tessellation A tessellation is an arrangement of shapes which fit together to fill a space with NO gaps or overlaps. *Originally the word was only applied to 2-dimensional space but is now taken to apply to 3-dimensional space as well. It must always be clear that the pattern used for filling can be continued indefinitely. It is usually assumed that the shapes used all have the same length of edge and touch each other along the full lengths of their adjoining edges, although in some ornamental work this is not the case.*

tiling A tiling is a **tessellation** in 2-dimensional space. *Additional requirements are sometimes applied such as: all the shapes must be identical; shapes may not be turned over; shapes may not be rotated; and so on. The individual shapes, or tiles, are usually polygons but curved shapes can also be used. Some of the more unusual shapes that can form the tile are: any* **pentomino**, *any* **hexomino**, *or any* **polyiamond** *made from 6 or 8 triangles.*

mosaic ≡ **tiling**

notation In a **tiling**, where 3 or more vertices meet at a common point, each separate vertex is given a number recording how many edges the shape to which that vertex belongs has. All the numbers around that common point, taken clockwise or anticlockwise in order, serve to record its type.
Example: At the common point marked in the centre of the diagram on the right, the numbers 3, 3, 4, 3, 4 would indicate that a triangle, a triangle, a square, a triangle, and a square meet at that point. This is usually abbreviated to 3^2, 4, 3, 4

regular tilings are those **tilings** made by using only a single shape which is a **regular polygon**. *There are only 3 possibilities: using an equilateral triangle, a square, or a regular hexagon.*

semi-regular tilings are those **tilings** which use more than one shape, but all the shapes are **regular polygons** and the **notation** at every point where vertices meet is the same. *There are only 8 possibilities, although one of them can exist in 2 forms with one being a mirror image of the other. The only regular polygons used are the triangle, square, hexagon, octagon and dodecagon.*

periodic tiling A tiling is said to be periodic if a section of it can be identified which will itself do the tiling by means of a translation only. *The section may be as large or as small as necessary – in effect it becomes the shape which is used to do the tiling. Most of the tilings which are seen are periodic. A tiling which not periodic is described as* **non-periodic** *or* **aperiodic**.

Maurits Escher (1898–1972) was a Dutch artist who produced a considerable amount of original art based on tilings. He fitted all sorts of shapes (real and imagined) together – birds, fish, insects and other animal-like objects.

rep-tile A rep-tile (≡ replicating tile) is a 2-dimensional shape of which multiple copies can be put together to make another shape which is **similar** to the original. *The simplest shapes with which this can be done are the right-angled isosceles triangle and an oblong (or parallelogram) whose edges are in the ratio $1:\sqrt{2}$ (which are the proportions used by A-sized papers). A more complex example is shown on the right. Many other rep-tiles can be found.*

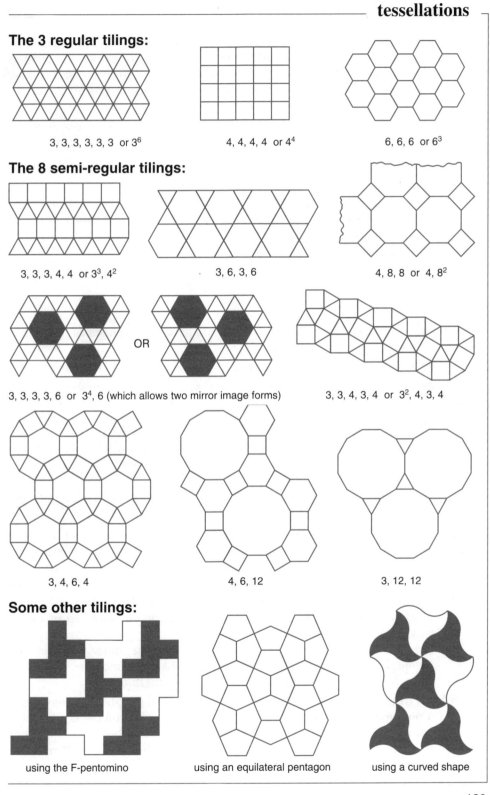

The 3 regular tilings:

3, 3, 3, 3, 3, 3 or 3^6

4, 4, 4, 4 or 4^4

6, 6, 6 or 6^3

The 8 semi-regular tilings:

3, 3, 3, 4, 4 or 3^3, 4^2

3, 6, 3, 6

4, 8, 8 or 4, 8^2

OR

3, 3, 3, 3, 6 or 3^4, 6 (which allows two mirror image forms)

3, 3, 4, 3, 4 or 3^2, 4, 3, 4

3, 4, 6, 4

4, 6, 12

3, 12, 12

Some other tilings:

using the F-pentomino

using an equilateral pentagon

using a curved shape

topology is the study of shapes and their properties which are not changed by transformations of a particular type. *It is popularly known as 'rubber-sheet' geometry because of the way shapes can be deformed, but it is more open than that because it is also possible to move lines from 'inside' a space to 'outside' working within the allowed rules.*

graph A topological graph is made up of a set of points and lines joining them.

vertex A vertex in a topological **graph** is one of the points which make the graph.

edge An edge in a topological **graph** is one of the lines which make the graph, and which must have a vertex at each end.

face A face in a topological **graph** is any single space completely enclosed by **edges**. *The space surrounding the graph, outside its boundary edges, is considered as one of the faces of that graph.*

network ≡ **graph** (of the topological variety)

node ≡ **vertex**

arc ≡ **edge**

region ≡ **face**

order of a vertex The order of a vertex is a number which states how many **edges** are joined to that **vertex**.

even vertex An even vertex is a **vertex** whose **order** is an EVEN number.

odd vertex An odd vertex is a **vertex** whose **order** is an ODD number.

Euler's formula states that in any topological **graph:**

> Number of faces + Number of vertices − Number of edges = 2

traversable A topological **graph** is said to be traversable if it can be drawn as one continuous line without going over any **edge** more than once. *Such a graph can only have either zero or two odd vertices.*

unicursal A topological **graph** is said to be unicursal if it is **traversable** and any start and finish are at the same point. *It will have NO odd vertices.*

topological transformations allow a shape to be deformed in almost any way provided that it always retains: the same number of **vertices**, **edges** and **faces**; the same **order** for all the vertices; the points along each edge following in the same relative positions. *Length and direction do not matter.*

topologically equivalent Two shapes are said to be topologically equivalent if, using only **topological transformations**, one can be deformed in such a way as to become identical to the other.
Example: Bus and rail organisations often use diagrams that are topologically equivalent to the real layout of the roads and rails, to simplify them. No measurements can be made on such diagrams.

Schlegel diagram A Schlegel diagram is a topological **graph** which represents a **polyhedron**. It is made by representing the polyhedron by its edges and deforming those, using only **topological transformations**, so that it lies flat. *The edges, vertices and faces of the polyhedron become those of the graph.*

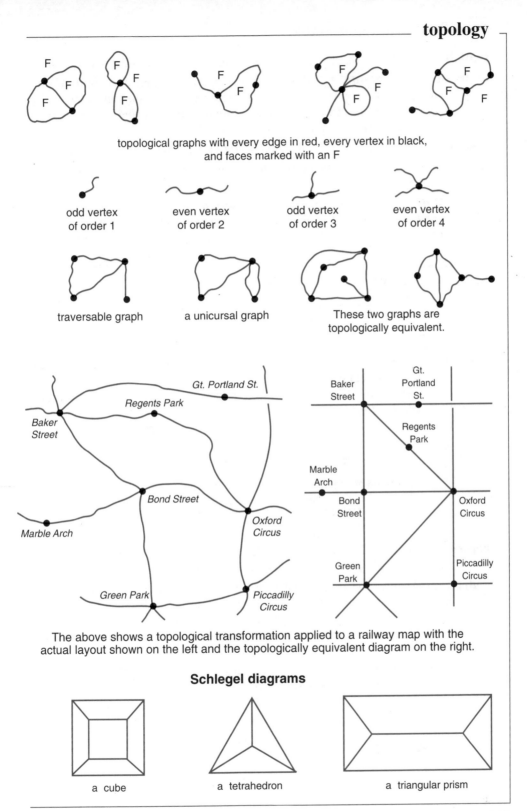

topological graphs with every edge in red, every vertex in black,
and faces marked with an F

odd vertex
of order 1

even vertex
of order 2

odd vertex
of order 3

even vertex
of order 4

traversable graph

a unicursal graph

These two graphs are
topologically equivalent.

Gt. Portland St.

Regents Park

Baker
Street

Bond Street

Oxford
Circus

Marble Arch

Green Park

Piccadilly
Circus

Baker
Street

Gt.
Portland
St.

Regents
Park

Marble
Arch

Bond
Street

Oxford
Circus

Green
Park

Piccadilly
Circus

The above shows a topological transformation applied to a railway map with the
actual layout shown on the left and the topologically equivalent diagram on the right.

Schlegel diagrams

a cube

a tetrahedron

a triangular prism

transformation A transformation is a change carried out under specific rules.

transformation geometry deals with the operations that may be used on a figure to affect its position, shape or size; or any combination of those. *The figure may be as small as a single point, and anything larger may be thought of as being made up of many points.*

object The object is the original shape BEFORE a **transformation** is applied.

image The image is the shape which appears AFTER the **transformation** has been applied to the **object**.

translation A translation is a **transformation** such that every point in the **object** can be joined to its corresponding point in the **image** by a set of straight lines which are all parallel and of equal length. *A translation is described by the direction and length of the movement.*
Examples:

a translation
of 3 cm right

a translation
of 1.5 cm up

rotation A rotation is a **transformation** about a fixed point such that every point in the **object** turns through the same angle relative to that fixed point. *A rotation is described by giving the angle and direction of the turn, and the position of the fixed point about which the turn is made.*
Examples:

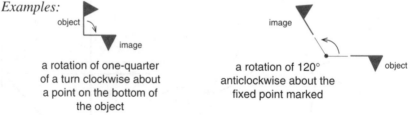

a rotation of one-quarter
of a turn clockwise about
a point on the bottom of
the object

a rotation of 120°
anticlockwise about the
fixed point marked

centre of rotation The centre of rotation is the fixed point about which the **rotation** takes place. *In the examples above each has a centre of rotation.*

reflection A reflection is a **transformation** such that any two corresponding points in the **object** and the **image** are both the same distance from a fixed straight line, and a line drawn between those points would be perpendicular to that fixed line. *It is described by giving the position of the fixed line.*

mirror line A mirror line is the fixed line used in making a **reflection**.
Examples: Each of these shows the effect of reflecting an object in the mirror line which is shown as a single black line. Note that, in this case, the words object and image could be interchanged.

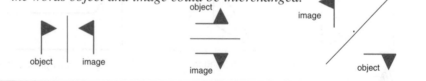

glide reflection A glide reflection is a **transformation** made by combining a **translation** with a **reflection** whose **mirror line** is parallel to the direction of the translation.
Example: A repeated glide reflection used to make a pattern.

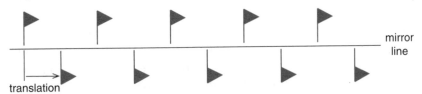

isometry An isometry is a **transformation** or combination of transformations, such that every distance measured between a pair of points in the **object** is the same as the distance between the corresponding pair of points in the **image**. *The object and image are the same shape and size. Translations, rotations, reflections and glide reflections are all isometries.*

direct isometry A direct isometry is an **isometry** in which either no **reflection**, or else an even number of them, has been used. *Using an odd number of reflections produces an* **opposite isometry**.

enlargement An enlargement is a **transformation** in which the distances between every pair of points in the **object** are multiplied by the same amount to produce the **image**. *The multiplier can take any value – a whole number or a fraction – but not zero.*
Examples:

scale factor A scale factor is the value of the multiplier used to make an **enlargement**. *Note that the scale factor is a multiplier for changing lengths only. The multiplier which affects the area will be (the scale factor)^2. For the change in volume of a 3-D shape it is (the scale factor)^3.*

centre of enlargement When the **object** and **image** of an **enlargement** have their corresponding points joined by straight lines, then all those lines will cross at a common point called the centre of enlargement.

shear A shear is a **transformation** in which all the lines in the **object** parallel to some fixed line (usually referred to as the base line) are moved in a direction parallel to that line, and an amount which is proportional to their distance from that line. *A shear is not an isometry, but the areas of the object and image are the same.*
Examples:

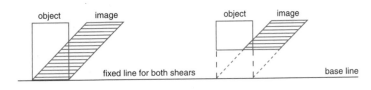

triangle   A triangle is a **polygon** which has 3 edges. *Its three interior vertex angles add up to 180 degrees. Triangles are usually described by reference to their edges or their vertices (or both).*

scalene triangle   A scalene triangle has ALL its edges of DIFFERENT lengths. *All of its vertex angles must also be of different sizes. It has no symmetry.*

isosceles triangle   An isosceles triangle has TWO edges of the SAME length. *Two of its vertex angles must also be of the same size. It has one line of symmetry.*

equilateral triangle   An equilateral triangle has ALL of its edges of the SAME length. *All its vertex angles are of the same size and equal to 60°. It has three lines of symmetry and rotational symmetry of order 3*

acute triangle   An acute triangle has NO vertex angle GREATER than 90°. *It must also be one of the above types of triangle.*

obtuse triangle   An obtuse triangle has ONE vertex angle GREATER than 90°. *It is also either an isosceles or a scalene triangle.*

right-angled triangle   A right-angled triangle has ONE vertex angle EQUAL to 90° *It is also either an isosceles or a scalene triangle.*

hypotenuse   The hypotenuse is the edge of a **right-angled triangle** which is opposite to the right angle. *It is also the longest edge of that triangle.*

base   The base of a **triangle** is any edge chosen to serve that purpose. *Usually it is the edge which is at the 'bottom' when the triangle is in a given position.*

perpendicular (height)   A perpendicular of a triangle is a line from a vertex to the opposite edge (extended if necessary) and at right angles to that edge; and the perpendicular height is the length of that line. *Any triangle has 3 perpendiculars, and these 3 (extended if necessary) all cross at the same point called the* **orthocentre**.

altitude ≡ **perpendicular height** or **perpendicular**.

median   The median of a triangle is a straight line joining one vertex of a **triangle** to the middle of the opposite edge. *Any triangle has three medians and they all cross each other at the same point.*

median triangle   The median triangle is the one formed by drawing straight lines between the mid-points of the three edges of another triangle. *A median triangle divides the original triangle into four congruent triangles.*

pedal triangle   The pedal triangle of a given triangle is the triangle formed by drawing straight lines between the points at the feet of the three altitudes of the original triangle. *It is also the triangle having the smallest possible perimeter that can be inscribed in the given triangle and touches all its edges.*

Euler points   The Euler points of a triangle are the mid-points of the three lines joining each vertex to the **orthocentre**.

centroid   The centroid of any triangle is the point at which the three **medians** cross. *The point is situated two-thirds of the length of each median from the corresponding vertex. If the triangle is actually made of some material of uniform thickness and density, the centroid is also the centre of mass.*

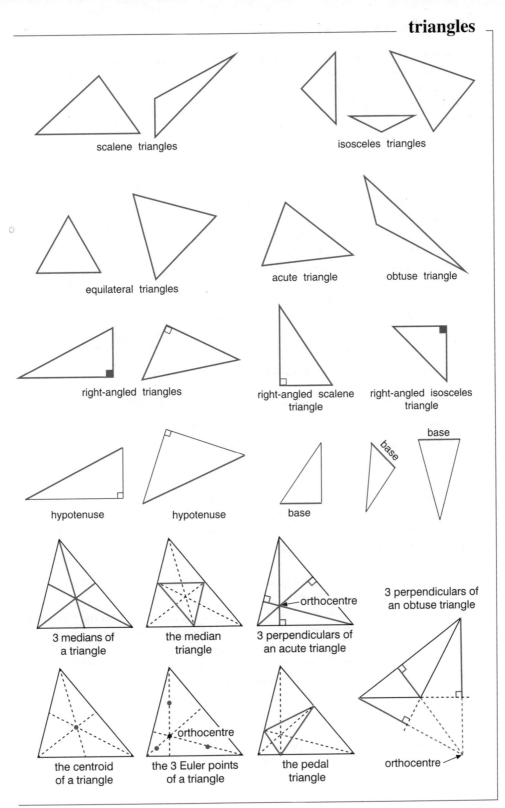

scalene triangles

isosceles triangles

equilateral triangles

acute triangle

obtuse triangle

right-angled triangles

right-angled scalene triangle

right-angled isosceles triangle

hypotenuse

hypotenuse

base

base

base

3 medians of a triangle

the median triangle

3 perpendiculars of an acute triangle

orthocentre

3 perpendiculars of an obtuse triangle

the centroid of a triangle

the 3 Euler points of a triangle

orthocentre

the pedal triangle

orthocentre

trigonometry (basic) ——————————————

trigonometry is the study of triangles with regard to their measurements and the relationships between those measurements using **trigonometric ratios**, and also goes on to deal with **trigonometric functions**.

notation Triangles are usually identified by 3 capital letters placed at each vertex (like ABC). There are then two methods for referring to the edges and the angles. One is descriptive, identifying edges by the two letters of the vertices at each end of that edge (AB, BC, AC) and the vertex angles by the three letters of the vertices to be taken in order as they trace out that angle (ABC, BCA, CAB). The other method, used in formulas, is to identify edges by a single small letter (a, b, c) corresponding to the letter of the opposite vertex, and the vertex angles by the single capital letter of that vertex (A, B, C).

trigonometric ratios express the relationship which exists between the size of one angle and the lengths of two edges in a right-angled triangle. *These ratios can be defined in relation to the triangle shown on the right where A, B and C are the 3 angles and a, b and c are the lengths of the 3 corresponding edges. There are 6 of these ratios:*

$$\textbf{sine } A = \frac{a}{c} \qquad \textbf{cosine } A = \frac{b}{c} \qquad \textbf{tangent } A = \frac{a}{b}$$

$$\textbf{cosecant } A = \frac{c}{a} \qquad \textbf{secant } A = \frac{c}{b} \qquad \textbf{cotangent } A = \frac{b}{a}$$

usually written as: sin A   cos A   tan A   cosec A   sec A   cot A

Some relationships between these ratios are:

$$\sin^2 A + \cos^2 A = 1 \qquad \tan A = \frac{\sin A}{\cos A}$$

$$\text{cosec } A = \frac{1}{\sin A} \qquad \sec A = \frac{1}{\cos A} \qquad \cot A = \frac{1}{\tan A}$$

In any of the fractions given above, none of the bottom values can be zero.

$\sin^{-1}$... ≡ **arcsin**... ≡ **inverse sine**... Any of these mean that the size of the angle which corresponds to the number given by ... has to be found. *Unless otherwise known, this is an angle between ⁻90° and ⁺90° Examples: $\sin^{-1}(0.5)$ is 30°   arcsin (⁻0.5) is ⁻30°*

$\cos^{-1}$... ≡ **arccos**... ≡ **inverse cosine**... Any of these mean that the size of the angle which corresponds to the number given by ... has to be found. *Unless otherwise known, this is an angle between 0° and ⁺180° Examples: arccos (0.5) is 60°   inverse cosine (⁻0.5) is 120°*

$\tan^{-1}$... ≡ **arctan**... ≡ **inverse tangent**... Any of these mean that the size of the angle which corresponds to the number given by ... has to be found. *Unless otherwise known, this is an angle between ⁻90° and ⁺90° Examples: inverse tangent (0.5) is 26.6°   arctan (⁻0.5) is ⁻26.6°*

complementary ratios are those **trigonometric ratios** which are connected by their **complementary angles**. *The three most important are:*
$$\sin \theta = \cos(90 - \theta) \qquad \cos \theta = \sin(90 - \theta) \qquad \tan \theta = \cot(90 - \theta)$$

Pythagoras' theorem In any right-angled triangle, the area of the square drawn on the **hypotenuse** (= *longest edge*) is equal to the total area of the squares drawn on the other two edges. *In the diagram, the area of square* Z *is equal to the areas of square* X *and square* Y *added together.* *In terms of the edge lengths a, b and c as shown in the right-angled triangle under* **trigonometric ratios***:*

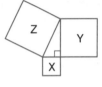

$$c^2 = a^2 + b^2 \quad or \quad a^2 = c^2 - b^2 \quad or \quad b^2 = c^2 - a^2$$
where c MUST be the hypotenuse.

sine rule The sine rule is based on the fact that, in any triangle the length of any edge is **proportional** to the **sine** of the angle opposite to that edge.

$$\frac{a}{\sin A} = \frac{b}{\sin B} = \frac{c}{\sin C}$$

cosine rule The cosine rule is an extension of **Pythagoras' theorem** which allows it to be applied to any triangle.

$$a^2 = b^2 + c^2 - 2bc \cos A$$
$$b^2 = c^2 + a^2 - 2ca \cos B \quad or \quad \cos A = \frac{b^2 + c^2 - a^2}{2bc}$$
$$c^2 = a^2 + b^2 - 2ab \cos C$$

cyclic formulas are **formulas** in which the letters identifying the various quantities can be systematically exchanged and keep the formula correct. In the **cosine rule** above, the three formulas for a^2, b^2 and c^2 are really the same formula with a, b, c changed around and the angle (A, B or C) altered to match. The formula for cos A on their right, is a **transposition** of the formula for a^2 and is also cyclic. In using cyclic formulas it is important to match the letters in the formula to the data for the triangle. Note the formula statement for **Pythagoras's theorem** is not cyclic since c must be the hypotenuse.

solving a triangle means that, from a given (limited) amount of information about a triangle, the remainder of its dimensions have to be found. Usually the given information, and what has to be found, concern only the lengths of the edges and the angles of the vertices.

ambiguous case In solving a triangle for which the lengths of two edges are known, and also the size of an angle which is NOT between those two edges, it is sometimes possible to find two values for the length of the third edge. *Example: Triangle* ABC *has* AB = 30; BC = 21; ∠BAC = 40° *Use of the* **sine rule** *gives sin* C = 30 sin 40° ÷ 21 = 0.918 *which means* ∠ACB *could be 66.7° or 113.3° so* ∠ABC *could be 73.3° or 26.7° and the corresponding lengths of* AC *could be 31.3 or 14.7 This is shown by marking* C' *and* C".

area This formula (which is **cyclic**) is for finding the area of a triangle when the lengths of 2 edges are known and also the size of the angle between them:

$$\text{Area} = \tfrac{1}{2} ab \sin C = \tfrac{1}{2} bc \sin A = \tfrac{1}{2} ca \sin B$$

trigonometry (further)

sine curve A sine curve is the graph showing how the value of the sine of an angle changes with the size of the angle. *The values have an upper bound of 1 and a lower bound of ⁻1.*

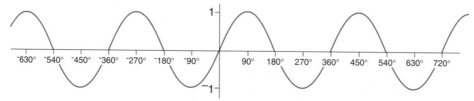

cosine curve A cosine curve is the graph showing how the value of the cosine of an angle changes with the size of the angle. *The values have an upper bound of 1 and a lower bound of ⁻1. It is the sine curve shifted left by 90°.*

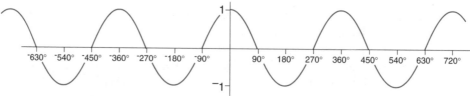

tangent curve A tangent curve is the graph showing how the value of the tangent of an angle changes with the size of the angle. *A tangent graph is drawn on the right.*

periodicity The periodicity of a curve is a measure of the distance a curve goes before it repeats itself. *The sine and cosine curves both have a periodicity of 360° while the tangent curve has a periodicity of 180°.*

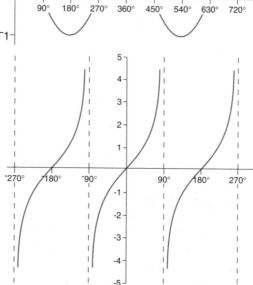

general angles Given only the value of a **trigonometric ratio** it can be seen, from the above graphs, that there are an infinite number of angles which can have that value. This is covered by the following three formulas:

If $\sin A = \sin \alpha$	then $A = 180n + (^-1)^n \alpha$		
If $\cos A = \cos \alpha$	then $A = 360n \pm \alpha$	where	n is any integer
If $\tan A = \tan \alpha$	then $A = 180n + \alpha$		

Example: $\sin A = 0.5$ (which is $\sin 30°$) so $A = 180 \times n + (^-1)^n \times 30$ Then A can be 30° or 150° or 390° or 510° or 750° or 870° ...

negative angles The **trigonometric ratios** of negative angles can be found using:

$$\sin(^-\theta) = {^-}\sin \theta \qquad \cos(^-\theta) = \cos \theta \qquad \tan(^-\theta) = {^-}\tan \theta$$

Example: $\sin(^-30°) = {^-}\sin 30° = {^-}0.5$

half-angle formulas for triangles are useful in solving triangles when the lengths of all three edges are known.

$$\sin\frac{A}{2} = \sqrt{\frac{(s-b)(s-c)}{bc}}$$

$$\cos\frac{A}{2} = \sqrt{\frac{s(s-a)}{bc}}$$

where

a, b, c, are the lengths of the edges
s is the length of the semi-perimeter
$s = (a + b + c) \div 2$

$$\tan\frac{A}{2} = \sqrt{\frac{(s-b)(s-c)}{s(s-a)}}$$

trigonometric formulas Apart from solving triangles, many formulas in trigonometry have been devised to allow expressions which involve the **trigonometric ratios** of two angles combined, to be rewritten in terms of the trigonometric ratios of one angle. The two main groups of these formulas deal with **addition** (or subtraction) and **multiple angles**. Care must be taken with these formulas that values are defined for the angles being used.

addition formulas

$\sin(A + B) = \sin A \cos B + \cos A \sin B$

$\sin(A - B) = \sin A \cos B - \cos A \sin B$

$\cos(A + B) = \cos A \cos B - \sin A \sin B$

$\cos(A - B) = \cos A \cos B + \sin A \sin B$

$\tan(A + B) = \dfrac{\tan A + \tan B}{(1 - \tan A \tan B)}$

$\tan(A - B) = \dfrac{\tan A - \tan B}{(1 - \tan A \tan B)}$

$\sin A + \sin B = 2\sin\left(\dfrac{A + B}{2}\right)\cos\left(\dfrac{A - B}{2}\right)$

$\sin A - \sin B = 2\cos\left(\dfrac{A + B}{2}\right)\sin\left(\dfrac{A - B}{2}\right)$

$\cos A + \cos B = 2\cos\left(\dfrac{A + B}{2}\right)\cos\left(\dfrac{A - B}{2}\right)$

$\cos A - \cos B = {}^-2\sin\left(\dfrac{A + B}{2}\right)\sin\left(\dfrac{A - B}{2}\right)$

multiple-angle formulas

$\sin 2A = 2\sin A \cos A$

$\cos 2A = \cos^2 A - \sin^2 A = 2\cos^2 A - 1$

$\tan 2A = \dfrac{2\tan A}{(1 - \tan^2 A)}$

$\sin 3A = 3\sin A - 4\sin^3 A$

$\cos 3A = 4\cos^3 A - 3\cos A$

trigonometric functions The trigonometic functions are the group of **functions** (linked to the **trigonometric ratios**) which map the size of any angle to a real number. *This is done by expressing each of the functions as an* **infinite series**. *The series for the sine, cosine and tangent functions are given below where the angle* **x** *has to be expressed in* **radians**. *The coefficients needed for sin **x** and cos **x** follow an obvious pattern. This is not so for tan **x** where they are generated by a more complex relationship.*

$$\sin x = x - \frac{x^3}{3!} + \frac{x^5}{5!} - \frac{x^7}{7!} + \frac{x^9}{9!} - \ldots$$

$$\cos x = 1 - \frac{x^2}{2!} + \frac{x^4}{4!} - \frac{x^6}{6!} + \frac{x^8}{8!} - \ldots$$

$$\tan x = x + \frac{x^3}{3} + \frac{2x^5}{15} + \frac{17x^7}{315} + \frac{62x^9}{2835} + \ldots$$

circular functions $\equiv$ **trigonometric functions**

units and conversions

There are three main systems of measurement still in use. These are known as: **imperial units,** shown here as (UK), American units shown as (US), and **metric units,** which form the basis of the Système Internationale.
The units in each of those systems are:

Length (UK and US)

12	inches	≡ 1 foot
3	feet	≡ 1 yard
22	yards	≡ 1 chain
10	chains	≡ 1 furlong
8	furlongs	≡ 1 mile
1760	yards	≡ 1 mile

Length (Metric)

10	millimetres	≡ 1 centimetre (cm)
10	centimetres	≡ 1 decimetre (dm)
10	decimetres	≡ 1 metre (m)
10	metres	≡ 1 decametre (dam)
10	decametres	≡ 1 hectometre (hm)
10	hectometres	≡ 1 kilometre (km)

Area (UK and US)

144	sq inches	≡ 1 square foot
9	sq feet	≡ 1 square yard
4840	sq yards	≡ 1 acre
640	acres	≡ 1 square mile

Area (Metric)

100	sq mm	≡ 1 sq cm (cm^2)
10000	sq cm	≡ 1 sq metre (m^2)
100	sq metres	≡ 1 are (a)
100	ares	≡ 1 hectare (ha)
100	hectares	≡ 1 sq kilometre (km^2)

Volume (UK and US)

| 1728 | cu. inches | ≡ 1 cubic foot |
| 27 | cu. feet | ≡ 1 cubic yard |

Volume (Metric)

1000	cu. mm	≡ 1 cu cm (cm^3)
1000	cu. cm	≡ 1 cu dm (dm^3)
1000	cu. dm	≡ 1 cu metre (m^3)

Mass (UK and US)

437.5	grains	≡ 1 ounce
16	ounces	≡ 1 pound
2000	pounds	≡ 1 short (US) ton
2240	pounds	≡ 1 long (UK) ton

Mass (Metric)

200	milligrams	≡ 1 carat
1000	milligrams (mg)	≡ 1 gram (g)
1000	grams	≡ 1 kilogram (kg)
1000	kilograms	≡ 1 tonne (t)

Capacity (UK liquid and dry, US liquid)

4	gills	≡ 1 pint
2	pints	≡ 1 quart
4	quarts	≡ 1 gallon

Capacity (US dry)

2	pints	≡ 1 quart
8	quarts	≡ 1 peck
4	pecks	≡ 1 bushel

Capacity (metric)

10	millilitres	≡ 1 centilitre (cl)
1000	millilitres	≡ 1 litre (l or L)
1000	litres	≡ 1 cu metre

Note that UK and US gallons are not the same size, so other measures of capacity having the same name are not the same size. Also a liquid pint is not the same size as a dry pint.

Time

60	seconds	≡ 1 minute
60	minutes	≡ 1 hour
24	hours	≡ 1 day
7	days	≡ 1 week
1	year	≡ 365 days
1	leap year	≡ 366 days

For the months:

30 days hath September,
April, June and November,
All the rest have 31,
Excepting February which,
Has 28 days clear,
But 29 in each Leap Year.

Time is based upon the second which is defined exactly. But a calendar needs to be based upon a year, which is the period taken by the Earth to go around the Sun. This takes 365.2425 days which, to give a workable system, means there must be a leap year every 4 years to adjust for that fraction of a day. That, by itself, gives a slight over-correction so, every 100 years another correction is made. This leads to the rule that :

a leap year is a year whose number can be divided exactly by 4, except that when it is the beginning of a century it must be divided by 400. So 1996 and 2000 are leap years, but 1995 and 1900 were not.

Originally every system had its own standard on which the other measures were based. Now the SI (or metric) standards are accepted worldwide and all other measures are defined in terms of that. These values are exact:

1 yard ≡ 0.9144 metres

1 pound ≡ 0.453 592 37 kilograms

1 gallon (UK) ≡ 4.54609 litres

1 gallon (US liquid) ≡ 3.785 411 784 litres

1 bushel (US dry) ≡ 35.239 070 166 88 litres

troy ounce The old measure of a troy ounce is still allowed for precious metals and stones.   1 troy ounce ≡ 31.103 476 8 grams (exactly).

conversion factors are multipliers (or dividers) which can be used to change a numerical measure in one type of unit into its **equivalent measure** in another type of unit. The table below gives several conversion factors.

To change ...	*into* ...	multiply by ...	To change ...	*into* ...	multiply by ...
acres	*hectares*	0.4047	litres	*gallons (US)*	0.2642
acres	*sq miles*	0.001563	litres	*pints (UK)*	1.760
barrels (oil)	*gallons (UK)*	34.97	metres	*yards*	1.0936
barrels (oil)	*gallons (US)*	42*	miles	*kilometres*	1.609344*
centimetres	*inches*	0.3937	millimetres	*inches*	0.03937
cubic cm	*cu inches*	0.06102	ounces	*grams*	28.35
cubic feet	*cubic metres*	0.0283	pints (UK)	*litres*	0.5683
cubic feet	*gallons (UK)*	6.229	pints (US liquid)	*litres*	0.4732
cubic feet	*gallons (US)*	7.481	pounds	*kilograms*	0.4536
cubic inches	*cu cm*	16.39	square cm	*sq inches*	0.1550
cubic inches	*litres*	0.01639	square feet	*sq metres*	0.0929
cubic metres	*cubic feet*	35.31	square inches	*square cm*	6.4516*
feet	*metres*	0.3048*	square km	*square miles*	0.3861
gallons (UK)	*gallons (US)*	1.2009	square metres	*sq yards*	1.196
gallons (UK)	*litres*	4.54609*	square miles	*acres*	640*
gallons (US)	*gallons (UK)*	0.8327	square miles	*sq km*	2.590
gallons (US)	*litres*	3.785	square yards	*sq metres*	0.8361
grams	*ounces*	0.03527	tonnes	*kilograms*	1000*
hectares	*acres*	2.471	tonnes	*tons (long/UK)*	0.9842
hectares	*sq metres*	10000*	tonnes	*tons (short/US)*	1.1023
inches	*centimetres*	2.54*	tons (long/UK)	*kilograms*	1016
kilograms	*pounds*	2.2046	tons (long/UK)	*tonnes*	1.016
kilograms	*tons (long/UK)*	0.000984	tons (short/ US)	*kilograms*	907.2
kilograms	*tons (short/US)*	0.001102	tons (short/US)	*tonnes*	0.9072
kilometres	*miles*	0.6214	yards	*metres*	0.9144*
litres	*cubic inches*	61.02			
litres	*gallons (UK)*	0.21997			

* indicates an **exact** figure; all others are approximations.

Some rough approximations for making comparisons are:

1 kg	is about	2.2 lb		1 lb	is just under	0.5 kg
1 kg	is about	35 ounces		1 ounce	is just under	30 grams
1 km	is about	0.6 mile		1 mile	is about	1.6 km
1 metre	is just over	3 feet		1 foot	is about	30 cm
1 cm	is about	0.4 inch		1 inch	is about	2.5 cm

units and the SI

The International System of Units (usually identified as SI) officially came into being as le Système International d'Unités at the 11th General Conference of Weights and Measures held in Paris in October 1960. There have been a few changes since then, such as the definition of a metre in 1983.

The SI defines 7 base units. Four of these with their abbreviations (in brackets) and definitions are:

metre (m) The metre is the unit of length. It is the distance light travels, in a vacuum, in $\frac{1}{299\,792\,458}$ th of a second.

kilogram (kg) The kilogram is the unit of mass. It is the mass of an international prototype in the form of a platinum-iridium cylinder kept at Sèvres in France. *It is now the only base unit still defined in terms of a material object, and also the only one with a prefix (kilo-) already in place.*

second (s) The second is the unit of time. It is the time taken for 9 192 631 770 periods of vibration of the caesium-133 atom to occur.

kelvin (K) The kelvin is the unit of thermodynamic temperature. It is $\frac{1}{273.16}$ th of the thermodynamic temperature of the triple point of water.
It is named after the Scottish mathematician and physicist William Thomson, 1st Lord Kelvin (1824–1907).

The other 3 base units are: the ampere (A) for measuring current; the mole (mol) for measuring amounts of a substance; the candela (cd) for measuring the intensity of light.

There are two supplementary units: the **radian** for plane angular measure, and the steradian for 'solid' or 3-dimensional angles.

All other units are derived from these. For example, the unit of force is the newton which, in terms of the base units, is metre kilogram/second^2 or m kg s^{-2}

There are rules for using the SI. Some of the more important are:
- A unit may take only one **prefix**.
  Example: Millimillimetre is incorrect and should be micrometre.
- To make numbers easier to read they may be divided into groups of three separated by spaces (or half-spaces) but NOT commas.
- Whole numbers should be separated from their decimal part by a comma. A point is acceptable but it must be placed on the line of the bottom edge of the number and not in a mid-way position.

convention for letters Many units are **eponyms** as they are named after a person closely associated with them (Newton, Watt, Pascal, etc). The convention for writing units which are eponyms is that when written in full their initial letter is in lower case, but their symbol or abbreviation is made with a capital letter. The one exception to this is the litre. It should be 'l' (el) but, as that could be confused with '1' (one), the capital letter 'L' is allowed. Some units and their abbreviations are:

becquerel	Bq	hertz	Hz	lux	lx
coulomb	C	joule	J	newton	N
farad	F	litre	l or L	pascal	Pa

prefixes A prefix is a group of letters placed at the beginning of a word to make a new word with a modified meaning. The SI allows other units to be created from the standard ones by using prefixes, which act as multipliers. This list gives those prefixes, the single letter or symbol to be used in the abbreviated form, and the multiplying factor they represent, both in index notation and (for most of them) in full. Note the difference between using capital letters and small letters.

prefix	symbol	factor	in full
yotta-	Y	$\times 10^{24}$	
zetta-	Z	$\times 10^{21}$	
exa-	E	$\times 10^{18}$	1 000 000 000 000 000 000
peta-	P	$\times 10^{15}$	1 000 000 000 000 000
tera-	T	$\times 10^{12}$	1 000 000 000 000
giga-	G	$\times 10^{9}$	1 000 000 000
mega-	M	$\times 10^{6}$	1 000 000
kilo-	k	$\times 10^{3}$	1 000
*hecto-	h	$\times 10^{2}$	100
*deca-	da	$\times 10^{1}$	10
		10^{0}	1
*deci-	d	$\times 10^{-1}$	0.1
*centi-	c	$\times 10^{-2}$	0.01
milli-	m	$\times 10^{-3}$	0.001
micro-	µ	$\times 10^{-6}$	0.000 001
nano-	n	$\times 10^{-9}$	0.000 000 001
pico-	p	$\times 10^{-12}$	0.000 000 000 001
femto-	f	$\times 10^{-15}$	0.000 000 000 000 001
atto-	a	$\times 10^{-18}$	0.000 000 000 000 000 001
zepto-	z	$\times 10^{-21}$	
yocto-	y	$\times 10^{-24}$	

* is a prefix not originally in the SI but was in the metric system and has remained in use because it has proved so convenient for everyday units.

Examples: *MW is megawatts or millions of watts.*
ns is nanoseconds or thousand-millionths of a second.

Some standard values given in SI units are:

nautical mile The international nautical mile is 1852 metres.

knot A knot is a speed of 1 nautical mile per hour.

gravitational acceleration The rate of acceleration on the Earth's surface due to gravity is 9.806 65 metres per second squared

speed of light The speed of light is 2.9979×10^{8} metres per second

light year A light year is a distance of 9.4605×10^{15} metres

vector A vector is something which can be defined by two quantities: its size and its direction.
Examples: Velocity is a vector since it is described by giving both the speed of an object and the direction in which the object is moving. Speed alone is not a vector. Force is also a vector.

AB and $\overrightarrow{AB}$ are two of the symbols used in diagrammatic and written work to refer to a **vector** whose size is represented by the distance between A and B and whose direction is from A to B. *The size is often referred to some scale. A single letter is also used (**A** or **a** or A etc.) but only after the vector has been defined in some way. In written work A̰ and a̰ are also used.*

scalar A scalar is a quantity which can be completely defined by a single number. *It may or it may not have units attached.*
Examples: Length, mass, speed and numbers are all scalar quantities.

plane vector A plane vector is a **vector** whose direction can be given solely by reference to two-dimensional space. *This might be done by using coordinates such as (x,y), or by using an angle such as a compass direction.*

position vector A position vector is a **vector** which starts at some known point, and its finishing point gives a position relative to that starting point. *Any coordinate system may be thought of as a vector system, where the origin is the starting point of a position vector which then goes to (or finishes at) the given position.*

free vector A free vector is a **vector** which does not have a defined starting point and so can be placed anywhere in space.

absolute value The absolute value of a **vector** is its size. *Direction is ignored.*

scalar multiplication Scalar multiplication of a **vector** is carried out by multiplying the size of a vector by a single number. *If the number used is negative, the direction of the vector will be reversed.*

unit vector A unit vector is a **vector** which is to be considered as being the unit of size from which other vectors are made by **scalar multiplication**.

negative vector The negative of a given **vector** is another vector that is the same in size but opposite in direction to the given vector.

orthogonal vectors are **vectors** whose directions are at right angles to each other.

vector addition Two or more **vectors** can be added by joining them together end to end, always so that their directions 'follow on' from each other, and the answer is the single vector which can be drawn from the 'start' point to the 'finish' point of the vectors. *The order in which they are joined does not matter. Subtraction is done by the addition of a negative vector.*

resultant A resultant is the **vector** produced by the addition and/or subtraction of two or more vectors. *It is the single vector which can replace all the other vectors and still produce the same result.*

vector triangle A vector triangle is made when 3 **vectors** are added together to form a triangle whose **resultant** is zero.

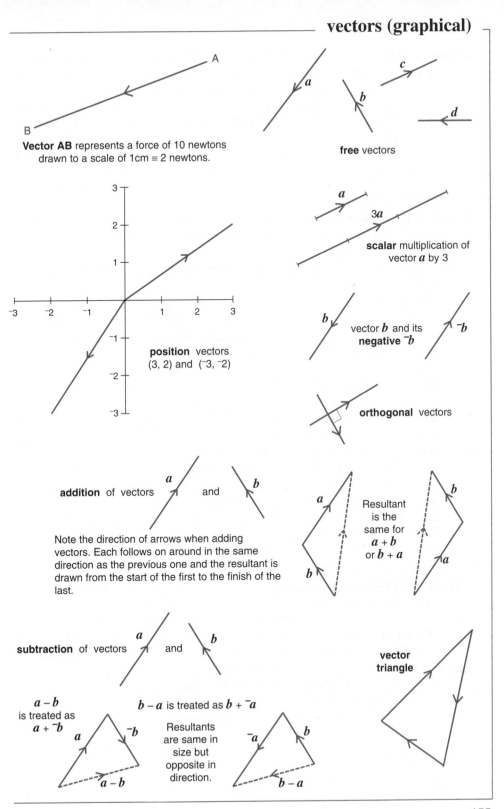

Vector AB represents a force of 10 newtons
drawn to a scale of 1cm ≡ 2 newtons.

free vectors

position vectors
(3, 2) and (⁻3, ⁻2)

scalar multiplication of
vector *a* by 3

vector *b* and its
negative ⁻*b*

orthogonal vectors

addition of vectors   *a*   and   *b*

Note the direction of arrows when adding
vectors. Each follows on around in the same
direction as the previous one and the resultant is
drawn from the start of the first to the finish of the
last.

Resultant
is the
same for
a + b
or *b + a*

subtraction of vectors   *a*   and   *b*

**vector
triangle**

a − b
is treated as
a + ⁻b

b − a is treated as *b + ⁻a*

Resultants
are same in
size but
opposite in
direction.

a − b

b − a

vector A vector is a set of numbers which can be represented in an appropriate space by a line to show both its length and its direction. *The line may be actual or imagined. The vector (2, 3) could be shown by an actual line drawn in a 2-dimensional space, while (1, 7, 5) would require a 3-dimensional space. With a vector having 4 numbers (6, 3, 8, 2) the space needed would be 4-dimensional and its representation could only be imagined. A vector having n numbers needs an n-dimensional space for its representation. Whilst a space having more than 3 dimensions cannot be easily visualised, it must be considered to exist if only in an abstract sense. Vectors are often written using angle brackets: ⟨7, 3, 14⟩. Commas are not always used.*

dimension The dimension of a **vector** is a measure of how many numbers are needed to express it, or the dimension of the space needed to show it.
Example: (1, 7, ⁻5) is a vector of dimension 3

row vector A row vector is a **vector** in which the set of numbers is written in a horizontal line.
Examples: (3, 5) (4, 0, 2) (15, ⁻6, 11, 28)   are all row vectors.
*Each can also be considered (without the commas) to be a **row matrix**.*

column vector A column vector is a **vector** in which the set of numbers is written in a vertical line.
*Examples are shown on the right. Each can also be considered to be a **column matrix**.*

$$\begin{pmatrix} 3 \\ 5 \end{pmatrix} \quad \begin{pmatrix} 4 \\ 10 \\ ^{-}2 \end{pmatrix} \quad \begin{pmatrix} 15 \\ 6 \\ 11 \\ 28 \end{pmatrix}$$

null vector A null vector is a **vector** which, when represented by a line, has no length and no direction; in its numerical form all the numbers will be 0. *It is also known as a **zero vector**.*

notation The most commonly used way of showing that vectors are being referred to in printed work is by using bold-face lower-case letters. *This does not define the vectors, but is the most economical way of handling them and, as with algebra, allows general statements to be made more conveniently.*
Example: Given $\mathbf{a} \equiv (7, 4, 12)$ and $\mathbf{b} \equiv (6, 0, 5)$ it is much easier to write $3\mathbf{a} + \mathbf{b}$ than to write the vectors out in full.
In written work, there are various ways of representing vectors.

scalar A scalar is any **real** number. *It is often shown by the symbol λ.*

scalar multiplication The result of the multiplication of a vector $\mathbf{v}$ by a **scalar** λ (written $\lambda\mathbf{v}$) where $\mathbf{v} \equiv (x_1 \quad x_2 \quad x_3 \quad x_4 \quad$ etc) is given by
$$\lambda\mathbf{v} \equiv (\lambda x_1 \quad \lambda x_2 \quad \lambda x_3 \quad \lambda x_4 \quad \text{etc })$$
The result is a vector which is λ times the size of the original. If λ is negative, the new vector will be in a reverse direction to the original.

unit vectors i, j, k In a 3-dimensional vector system, with the axes perpendicular to each other, the 3 unit vectors (length = 1) lie on the main axes (x, y, z) and are identified as **i j k** respectively. *Expressed numerically $\mathbf{i} \equiv (1, 0, 0)$, $\mathbf{j} \equiv (0, 1, 0)$ and $\mathbf{k} \equiv (0, 0, 1)$.*
All the other vectors can then be written in terms of the 3 unit vectors.
*Example: (4, 17, 9) ≡ (4**i**, 17**j**, 9**k**)*

magnitude The magnitude of a vector is a measure of its length when it is represented by a single straight line and is the positive square root of the **scalar product** of the vector with itself. *It is also known as the* **modulus**.

$|v|$ is the symbol for the **magnitude** of the vector shown. (*In this case* **v**).
   Example: Given **v** $\equiv$ (3, 1, 8) *then* $|v| = \sqrt{v.v} = \sqrt{3^2 + 1^2 + 8^2} = \sqrt{74}$

v is another symbol for the **magnitude** of the vector **v** where the letter of the vector is printed in italic and not in bold.

scalar product The scalar product of two vectors **a** and **b** (written **a.b**) is found by multiplying together their **magnitudes** and the cosine of the angle between them. *The result is a* **scalar**. *It is also known as the* **inner product** *or the* **dot product**. *It must not be confused with* **scalar multiplication**.

a.b = $|a| \times |b| \times \cos \theta$ more usually written as $|a| \, |b| \cos \theta$

In the case where the vectors are given in terms of a coordinate system where **a** $\equiv (x_1 \, x_2 \, x_3)$ and **b** $\equiv (y_1 \, y_2 \, y_3)$
       then **a.b** = $x_1 y_1 + x_2 y_2 + x_3 y_3$
Example: Given **a** $\equiv$ (3, 6, 5) and **b** $\equiv$ (4, 7, 2)
       then **a.b** = $3 \times 4 + 6 \times 7 + 5 \times 2 = 64$

For unit vectors: **i.j** = **j.k** = **k.i** = 0   and   **i.i** = **j.j** = **k.k** = 1

perpendicular vectors Two vectors must be perpendicular to each other if their **scalar product** is zero. *This can happen because* $\cos \theta = 0$ *or the product of their magnitudes is 0. They are also known as* **orthogonal vectors**. *Example: Given* **a** $\equiv$ (2, 7, 3) *and* **b** $\equiv$ (4, 1, $^-$5) *then since* $(2 \times 4 + 7 \times 1 + 3 \times {}^-5) = 0$, *they must be perpendicular to each other.*

right-handed system Three vectors **a**, **b**, **c** (in that order, not all in the same plane, and all starting at O) form a right-handed system if, when looking in the direction of **c**, **a** can be rotated CLOCKWISE about O through an angle of less than 180° to lie on **b**. *If the required rotation (which must be <180°) is* ANTICLOCKWISE *then it is a* **left-handed** *system. If the vectors* **a**, **b**, **c** *form a right-handed system then so also do* **b**, **c**, **a** *and* **c**, **a**, **b**. *A right-handed system becomes left-handed if one vector is reversed. The unit vector system* **i**, **j**, **k** *is always a right-handed system.*

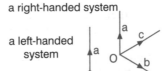

a right-handed system

a left-handed system

vector product The vector product of two vectors **a** and **b** (written **a** × **b** or **a** ∧ **b**) with an angle between them of θ ($0 < \theta < 180°$) is vector **c** whose **magnitude** is the product of the magnitudes of **a**, **b** and $\sin \theta$ and is perpendicular to both **a** and **b** in a direction that makes **a**, **b**, **c** a **right-handed system**.

a × **b** = $|a| \times |b| \times \sin \theta$ more usually written as $|a| \, |b| \sin \theta$

For unit vectors:

i × **j** = **k** **j** × **k** = **i** **k** × **i** = **j** and **i** × **i** = **j** × **j** = **k** × **k** = 0

parallel vectors Two vectors (neither of which is a zero vector) must be parallel to each other if their **vector product** is zero.

There are several pairs of words in mathematics which seem to be alike, either in look or meaning. Some are given in their appropriate section, but some are put together in this section with help to distinguish between them.

approximation These two are often used as being identical in meaning
estimation   because an estimation is an approximation based upon a judgement. However, an approximation is not necessarily an estimation. An **approximation** is applied to a number that already exists; an **estimation** creates a number by making a judgement that is usually (but not always) derived from an actual physical situation.

capacity Mathematically these both measure the size of a 3-dimensional space
volume   and use the same units. It is in their usage that they differ: capacity refers to a containing space and the room available to hold something; volume is the space actually occupied by an object or the bulk of some substance.
*Example: A bucket has a **capacity** of 20 litres so the **volume** of water needed to fill the bucket is 20 litres.*

mass   These two words are commonly used as though they were identical.
weight   This does not matter for ordinary use, but there is a difference. The **mass** of a body is a measure of how much matter it contains and is the property of that body that governs the way it will behave under the action of a force. The **weight** of a body is a measure of the force the body itself produces when in a gravitational field. The mass of a body is unchanged wherever it is in the Universe, whereas the weight depends upon the gravitational force at the place it is being weighed. Objects do not change their mass, but certainly weigh less, when on the Moon than when they are on the Earth. *Weight is a vector quantity.*

inclusive These words, used to indicate whether something is to be put in
exclusive (included) or not (excluded), are often used in counting situations, especially with reference to the calendar.
Examples: 'The insurance cover runs from the 5 August 1996 to 4 August 1997 inclusive'; so both the given dates are covered. 'The sale will run between 3 and 9 July exclusive'; so the sale is on for 5 days since the two given dates are not to be counted in.

bar chart These two types of frequency diagram look very similar and are
histogram only differentiated by the way the frequency is represented: by the LENGTH of the bars in the **bar chart** and by the AREA of the bars in the **histogram**. Bar charts tend to be used for **discrete data** while histograms are nearly always concerned with **continuous data**.

complementary Both of these, when referred to angles, are related
supplementary   to the right angle. To complement something is to complete it, and so **complementary** angles together make a complete right angle. To supplement something is to add something extra, in this case another right angle, so **supplementary** angles together make two right angles or 180°.

congruent These two words are used to compare geometrical shapes.
similar **Congruent** is much the stronger as the two shapes being compared must be identical to each other in every way, except that one may be turned around, or over, in relation to the other. **Similar** only requires the two shapes to be identical in their shape, and one may be bigger than the other. *Example: Given one large and one small square, they are similar but not congruent.*

conjecture The difference between these two words is mainly one of usage.
hypothesis Both are unproven statements for which, usually, a lot of supporting evidence can be found. **Conjectures** are generally about numbers and allow NO exceptions. **Hypotheses** generally occur in statistics, are usually set up before carrying out a search for supporting evidence, often involve a probability as to their correctness, and are sometimes put in a negative form as it is then easier to search for evidence that they are wrong.

possible An event is described as possible if it is within the bounds of reason
probable that it could happen; otherwise it is impossible. It is described as probable if the chance of it happening is a good one. Though 'good' is not defined, common usage implies that a probable event is one that is more likely to happen than not.
Example: Someone could say about a racehorse that is not very good: 'It is possible for this horse to win, but not probable.'

necessary Consider two related statements P and Q ordered P first, Q
sufficient second. If, whenever P is true then Q must also be true, then P is a **sufficient** condition for Q. If, whenever P is false then Q must also be false, then P is a **necessary** condition for Q.
Example: Referring to a quadrilateral, the following statements are made-
   Q *is 'It is a parallelogram'*
   P' *is 'Two opposite edges are equal in length'*
   P'' *is 'All its edges are equal in length'*
   P''' *is 'Two opposite edges are parallel and equal in length'*
Then, in relation to statement Q:
   P' *is necessary;* P'' *is sufficient:* P''' *is necessary and sufficient*

partition Both of these words describe a type of division applied to a
quotition physical situation, where whatever is being divided is named, or has units attached. If the **dividend** and **divisor** have DIFFERENT types of names it is a partition; if they have the SAME type of name it is a quotition.
Examples: Dividing (sharing) 100 apples among 10 people is a partition (apples ÷ people). Finding how many 10 cm lengths can be cut from a 1 metre strip is a quotition (length ÷ length).

disc There is no difference in the meanings of these, only in the spelling.
disk Currently there is a difference in usage with dis**k** being used for a computer disk and dis**c** for a compact disc. *The 'k' usage will probably win.*

word origins

Many of the words used in English (especially those in mathematics) were originally taken from, or based upon, the ancient languages of Greek and Latin. Some of these connections can be seen from this table.

Meaning	Greek	Latin	Used in
something learned	mathema		mathematics
small stone		calculus	calculate, calculus
to be worth		valere	value, evaluate
number	arithmos	numerus	arithmetic
one	hen	unus	unit
two	di	bi	diagonal, bisect
three	treis	tres	triangle, trisect
four	tetra	quadri	tetrahedron, quadrilateral
five	pente	quinque	pentagon, quintic
six	hexe	sex	hexagon, sextant
seven	hepta	septem	heptagon, September
eight	okta	octo	octagon, October
nine	ennea	nonus	enneagon, nonagon
ten	deka	decem	decagon, December
twelve	dodeka	duodecem	dodecagon, duodecimal
hundred	hekaton	centum	hectare, per cent
thousand	khilioi	mille	kilogram, per mil
many	polion	polium	polygon, polyhedron
one tenth		decimus	decimal
first	protos	primo	prototype, prime
single	monos	singularus	monopoly, singular
double		duplicatus	duplicate
half	hemi	semi	hemisphere, semicircle
corner	gonos	angulus	hexagon, angle
small circus	kirkos	circulus	circle
measure across	diametros	diametrus	diameter
round		circum	circumference
string	khorde	chorda	chord
to cut	secare		segment, secant
wheel		rota	rotate
roller	kulindros	cylindrus	cylinder
coil	speira	spira	spiral
done with care		accuratus	accurate
go astray		errare	error
land measuring	geometria		geometry
equal legs	isoskeles		isosceles triangle
unequal	skalenos		scalene triangle

Meaning	Greek	Latin	Used in
lying near		adjacere	adjacent
sharp		acutus	acute angle
blunt		obtusus	obtuse angle
maker		factor	factor
crowd		frequentia	frequency
thing given		datum	data
middle		medius	median
breaking		fractio	fraction
to name		denominare	denominator
to touch		tangere	tangent
heart-shaped	kardioeides		cardioid
kidney	nephros		nephroid
ball or globe	sphaira		sphere
sawn across	prisma		prism
separate	horizein		horizon
follow		sequi	sequence
join together		serere	series
foundation	hupothesis		hypothesis
put together		conjicere	conjecture
speculation	theorema		theorem
mark or token	sumbolon		symbol
a step		gradus	gradient
dwarf	nanos	nanus	nano-second
great	megas		mega-hertz
giant	gigas		giga-joules

Plurals

A result of using words from other languages is in the way plurals can be formed. Many are now seen with the -s or -es ending of standard English, but the following list gives the singular and plural forms of some words in mathematics which can take a different ending.

abscissa	abscissae	frustum	frusta	minimum	minima
apex	apices	helix	helices	parabola	parabolae
axis	axes	hyperbola	hyperbolae	polygon	polygona
datum	data*	hypothesis	hypotheses	polyhedron	polyhedra
die	dice	index	indices	radius	radii
directrix	directrices	locus	loci	rhombus	rhombi
focus	foci	matrix	matrices	trapezium	trapezia
formula	formulae	maximum	maxima	vertex	vertices

*Data, when used in statistics and computing, is increasingly used only as a singular collective noun – 'the data is' rather than 'the data are'.

words in further mathematics

number theory is a branch of mathematics which is concerned only with the study of numbers. *This covers matters such as types of numbers (primes, squares etc.) and their occurrence and relationship to each other, how number systems work, the effect of using different bases, and the development and improvement of arithmetic processes.* **Modulo arithmetic** *is a much used tool in this work. Mainly it is the positive whole numbers which are dealt with, but work is often extended to other sorts (negatives, irrationals etc.) and algebraic ideas are frequently used to find and prove many of the results. Most mathematicians have, at some time in their careers, turned their attention to number theory.*

coprime or **relatively prime** Two (or more) whole numbers are said to be coprime, or relatively prime, to each other if they have no **factors** (other than 1) in common.
Examples: 12 and 25 are coprime. So also are 8, 9 and 11

logarithms The logarithm of a number (N) to a particular base (b) is the **power** to which that base must be raised to equal that number. *This can be shown as:* $(base)^{logarithm\ of\ the\ Number} = the\ Number;\ or\ N = b^{log\ N}$
The importance of logarithms was in the way they could be used to simplify multiplication and division by using the **laws of indices***.*
Example: Given $M = b^{log\ M}$ *and* $N = b^{log\ N}$ *then*
$$M \times N = b^{log\ M} \times b^{log\ N} = b^{log\ M + log\ N}\ and\ the\ multiplication\ sum$$
has been replaced by an addition sum. There was a need to be able to find the logarithm of any number, and books of tables were provided for that.

common logarithms are calculated to a base value of 10 and were the **logarithms** once used very widely for doing arithmetical calculations. *The coming of the electronic calculator has removed the need for these.*

natural logarithms are calculated to a base value of e ($\approx 2.718\,28$), and arise in work involving the ideas of the **calculus**. *This is now the principal use for logarithms rather than the simplification of arithmetic.*

prismatoid Like a **prism** and a **prismoid** the prismatoid has two parallel end faces which are polygons but, in this case, these two polygons do not have to have the same number of vertices. The faces connecting the two ends must be either triangles or quadrilaterals, with all vertices being coincident with one or other of the vertices of the two end faces.

spherical trigonometry is the study of the measurements, and relationships between those measurements, of the triangles which can be drawn upon the surface of a sphere. *This is needed in making calculations relating to astronomical observations and navigation.*

versine is an abbreviation of versed sine.

versine θ (where θ is any angle) = 1 − cosine θ

Example: versine 30° = 1 − cosine 30° = 1 − 0.866 = 0.134 (to 3 d.p.)

haversine is an abbreviation of half-**versine**.

haversine θ = (versine θ) ÷ 2 or 0.5 × (1 − cosine θ)

This function is used in calculations in **spherical trigonometry**.

solid angle In 3-dimensional space a solid angle is formed by all those lines which start at a common point and pass through a simple closed curve. *Solid angles are to be found at the vertices of any 3-dimensional shape.*

steradian A steradian is the unit of measure of a **solid angle**. It is the size of the solid angle formed at the centre of a sphere of unit radius, by those lines radiating from the centre which cut off a **segment** whose curved surface is of unit area. *The maximum size of a solid angle is 4π steradians.*

bicimal is a shortened form of bicimal fraction. These are similar in structure to **decimal fractions** but the numbers are written in **base** 2 or **binary** form. *Example: The bicimal 0.101_2 is equivalent to the common fraction $\frac{101}{1000}$ when both the top and bottom numbers are in binary.* *Changing to base 10: $101_2 \rightarrow 5$ and $1000_2 \rightarrow 8$ so $\frac{101}{1000} \rightarrow \frac{5}{8}$ or 0.675* *This idea can be extended to other bases. The **ternary** system (base 3) would produce **tercimals**, and so on.*

cubic A cubic equation, expression or function is one of **degree** 3 *Example: $2x^3 + 4x^2 - 8 = 0$ is a cubic equation.*

quartic A quartic equation, expression or function is one of **degree** 4 *It is also known as a **biquadratic**.* *Example: $3x^4 - 5x^2 + 7$ is a quartic or biquadratic expression.*

quintic A quintic equation, expression or function is one of **degree** 5 *Example: $f(x) \equiv 3x^5 - 7x^2 + 4x$ is a quintic function.* *The general equation of a quintic (or higher degree) cannot be solved by use of straightforward formulas, unlike those of degree 4 or less.*

algebraic number An algebraic number is a **real number** which can arise as the root of a **polynomial equation** whose coefficients are all **whole numbers**. *All rational numbers are algebraic; but only some irrational numbers are.* *Example: $\sqrt{3}$ is an algebraic number since $\sqrt{3}$ is a root of $x^2 = 3$* *Any irrational number which is not algebraic is known as a **transcendental number**. The best known transcendental numbers are π and e.*

geometric mean The geometric mean of a set of n positive numbers is found by multiplying them all together and taking the positive nth root of the result. *Example: The geometric mean of $\{2,3\}$ is $\sqrt{6}$; of $\{4, 5, 6\}$ is $\sqrt[3]{120}$.*

derangements Given n different objects to be put into n boxes then there are $n!$ **(factorial n)** ways of doing it. However, if each object has to be matched to a particular box (say objects and boxes are numbered) then there can only be one way of doing it correctly. If EVERY object is in a WRONG box then that is known as a derangement of the objects. *(Some right and some wrong is not a derangement.)* *For any value of n the number of all possible derangements is shown by $!n$. Some values of $!n$ (also known as **subfactorial n**) are given in the table.*

n	$!n$
1	0
2	1
3	2
4	9
5	44
6	256
7	1854
8	14833

Example: Given 1, 2, 3 as the objects; then 2,3,1 and 3,1,2 are the only 2 possible derangements (1,3,2; 2,1,3; 3,2,1 are not) and so $!3$ is 2

curve In most cases curve is used to describe a line which is not straight. However, its more general use in mathematics requires that straight lines be included. *Example: 'Join the points with a curve' allows a straight line to be drawn if that is appropriate.*

clockwise The direction of a movement around a circular arc is described as being clockwise if it moves in the same direction, relative to the centre, as that of the hands of a conventional clock.

anticlockwise An anticlockwise direction is one which is opposite to **clockwise**.

certain An event can only be described as certain if it MUST happen.

likely In ordinary use an event is described as being likely if it has a better chance of happening than NOT happening. In mathematics, likely is more often used as having the same meaning as **probable**.
Examples: 'It is likely I shall go to town tomorrow.'
'How likely is it that I shall throw a 4 with this die?'

difference In ordinary use the difference between two things requires a description of the ways in which they are not alike, but, in arithmetic, a difference is the result of subtracting one number from another.
Example: The difference between 4 and 6 could be described as being 'one is all straight lines while the other is curved', but arithmetically it is 2

prefixes A prefix is the first part of a word that can be changed, as necessary, to adjust the meaning of the complete word. *The second part of the word, following the prefix, is called the 'stem'. A common use of prefix + stem in mathematics is in naming shapes, and for this purpose both of them are based on either Greek or Latin. The table gives the prefixes used to cover the meanings from 3 to 12. The stems used are:*

No.	Latin	Greek
3	tri-	tri-
4	quad-	tetra
5	quin-	pent-
6	sex-	hex-
7	sept-	hept-
8	oct(o)-	oct(a)-
9	nona-	ennea-
10	decem-	deca-
11	undeca-	hendeca-
12	duodecim-	dodeca-

Latin: -angle (= angle)   -lateral (= edge or side)
Greek: -gon (= angle)   -hedron (= base or face)
Examples: triangle, quadrilateral, pentagon, hexagon, etc.
The general rule is that both prefix and stem should come from the same language but two notable exceptions are nonagon (rather than enneagon, nonangle, or nonalateral), and undecagon (rather than hendecagon).
Usage over a long period of time has fixed the names we now use for 2- and 3-dimensional shapes as being mainly from the Greek, except for triangle (rather than trigon) and quadrilateral (rather than tetragon)
*The **SI** also makes extensive use of a different set of prefixes to show numbers, and many other prefixes can be found under **word origins**.*

magnitude The magnitude of something is a measure of its size. *It is generally used in relation to some scale so that comparisons can be made easily. Earthquakes and the brightness of stars use such scales. In mathematics* **standard form** *serves to do the same job. It is easy to see 2.45×10^{15} is bigger than 9.67×10^{14} but not when they are written out in full.*

finite A quantity is said to be finite if it can be counted or measured in some way. *The numbers involved might be enormous, and physically extremely difficult to count, but, as long as it is known or believed that whatever is being counted does not go on endlessly, then it is a finite quantity. The number of leaves on all the trees in the world, and the number of the atoms in the Universe are both finite quantities.*

infinite A quantity is said to be infinite if it clearly has some size but it cannot be counted or measured in any way. *While the word is usually applied to the very large it is also used in the phrase 'infinitely small'.*

evaluate To evaluate something is to work out its numerical value.
Example: 'Evaluate $x^2 + 5$ when $x = 3$' requires the answer 14

equivalent Two things are said to be equivalent to one another if they have the same value, or produce the same effect in use, but have different forms.
Example: £1 is equivalent (in value) to 100 pence.

property A property of an object or, more usually a group of objects, is some particular fact which is true for that object or, all the objects in the group.
Example: It is a property of squares that their diagonals cross at right angles.

invariable Any property of an object which never changes is said to be invariable.
Example: With a shear transformation, the area of a shape is invariable.

proportion One set of quantities is said to be in proportion to another set if a **mapping** between the two sets can be made EITHER by using a constant multiplier, OR by matching the pairs from each set so that their **product** is always the same constant value.

direct proportion When two sets are in **proportion** using a constant multiplier (so that an INCREASE in one matches an INCREASE in the other) they are said to be in direct proportion. *It is also known as **direct variation**.*
Example: The set {12, 20, 32} is in direct proportion to {3, 5, 8} and the constant multiplier is 4. Even when interchanged the two sets are still in direct proportion but the constant is now $\frac{1}{4}$

inverse proportion When two sets are in **proportion** and an INCREASE in one matches a DECREASE in the other they are said to be in inverse proportion. *It is also known as **indirect variation**.*
Example: The sets {40, 24, 15} and {3, 5, 8} are in inverse proportion to each other since $40 \times 3 = 120$, $24 \times 5 = 120$ and $15 \times 8 = 120$

∝ is the symbol for 'is **proportional** to' and k is used for the constant.
Examples: $y \propto x$ means y is in DIRECT proportion to x; or $y = kx$
$y \propto \frac{1}{x}$ means y is in INVERSE proportion to x; or $y = \frac{k}{x}$
where k is a numerical constant.

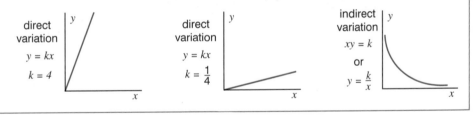

words of wider interest

ogee An ogee is a shallow S-shaped curve which is very often seen in the ornamentation used in the arches and ceilings of buildings, especially in those of earlier (classical) times. *It is also called a* **sigmoid**.

ogive An ogive is the name sometimes given to a **cumulative frequency diagram** where the curve is usually in the shape of an **ogee**.

minuend In a subtraction sum the minuend is the number from which the other number must be subtracted.

       minuend
  – subtrahend
    difference

subtrahend In a subtraction sum the subtrahend is the number which must be subtracted from the other number.

difference In a subtraction sum the answer is known as the difference.

addend In an addition sum each of the numbers is an addend. The first of the numbers is also known as the **augend**.

multiplicand In a multiplication sum the multiplicand is the first of the two numbers given.

  multiplicand
 × multiplier
   product

multiplier In a multiplication sum the multiplier is the second of the two numbers given. *Since multiplication is* **commutative** *and order does not matter, the multiplicand and multiplier can be changed around, so the names are only for identifying them in a particular case.*

aliquot part One number (or quantity) is an aliquot part of another if it divides into it exactly. *Usually this is applied only to whole numbers, and an aliquot part is the same as a* **proper factor**. *If it is not an aliquot part, then it can be referred to as an* **aliquant part** *though this term is rarely used. Examples: 3 is an aliquot part of 12; 7 is an aliquant part of 12*
          10p, 20p, 25p and 50p are all aliquot parts of £1; 75p is not.

pyramid number A pyramid number is made by adding together a consecutive set of **polygon number**s, starting at 1 (all of them being the same shape). *Example: Using the triangle number series (1, 3, 6, 10, 15, 21 etc.), the pyramid numbers 1, 4(=1 + 3), 10(=1 + 3 + 6), 20, 35, 56 etc. can be made. Pyramid numbers are identified by the shape of the polygon number used to make them, so the above sequence (1, 4, 10, 20, 35, 56 etc.) would be called triangular pyramid numbers. (In this particular case they could also be called tetrahedron numbers.) The first eight numbers of four types are:*

triangular pyramid	1	4	10	20	35	56	84	120
square pyramid	1	5	14	30	55	91	140	204
pentagonal pyramid	1	6	18	40	75	126	196	288
hexagonal pyramid	1	7	22	50	95	161	252	372

In any sequence of pyramid numbers based on a polygon having e edges, the nth number of the sequence is given by
$$n(n + 1)[e(n - 1) - 2n + 5] \div 6$$

Their name implies that, given a matching number of balls, they could be stacked into the shape of a pyramid. In fact, a good fit between the layers is only possible for the triangular and square types. No other orderly stacking is possible. A plan view of a square pyramid is shown, the layers going: 1 + 4 + 9 + 16 = 30

infix notation is the method used in writing mathematical expressions where the **operators** are written between the two numbers (or expressions) on which they have to operate.
Example: 1 + 2;   3 × 4;   (5 + 6) × (7 − 8) and so on.
This method offers scope for errors since the operators have to be used in a particular order, and not merely read from left to right.
Example: × has to be done before + so, 2 + 3 × 4 equals 14 not 20
To get the second answer requires brackets to be used: (2 + 3) × 4 = 20

reverse Polish notation was designed to remove all possible ambiguities of the **infix notation** by putting each operator immediately after the pair of numbers on which it must operate. *Brackets are not needed.*
Examples: 2 + 3 × 4 becomes 2 3 4 × + meaning first 3 × 4 then + 2
(2 + 3) × 4 becomes 4 2 3 + × meaning first 2 + 3 then × 4
Those expressions are simple, more complicated ones are harder to follow.
Example: 2 3 4 + × 5 6 × 7 + × should produce the answer 518
The 'trick' is to take each operator in turn (working from left to right), use it on the two preceding numbers and replace them (and the operator) with the result at each stage. The method is useful in computer programming. Some of the early electronic calculators required that the expression to be evaluated was input in this form, but these proved so unpopular they were soon discontinued. It is also known as **postfix notation**.

vinculum A vinculum is a horizontal bar placed over the top of a mathematical expression to indicate that it is to be treated together as one unit. *It is rarely seen (on its own) nowadays, having been replaced by the use of brackets.*
Example: 3x + 5 $\overline{x − y}$ would now be written 3x + 5(x − y)
It still appears in mathematics as a part of the square $\overline{root}$ symbol.
Example: √ $\overline{8x + y}$ is now written as √(8x + y) or √8x + y

quincunx A quincunx is an arrangement of 5 objects in a square, so that there is 1 in each corner and 1 in the middle. *It is commonly seen in the markings on dice, dominoes and playing cards.*

quinquangle A quinquangle is another name for a **pentagon**. *The word is seldom used now, but does serve as a quaint reminder of the place of Latin in our mathematical language.*

trivium In the Middle Ages (c.500–1500 AD) the trivium was that part of educational studies, made up of the three subjects: grammar, rhetoric and logic. *Rhetoric is the skill of producing convincing arguments.*

quadrivium In the Middle Ages the quadrivium was that part of educational studies, made up of the four subjects: arithmetic, geometry, astronomy and music. *These, together with the three subjects of the* **trivium** *were the seven liberal arts which formed the basis of most university courses in those days.*

quadrature The quadrature of a shape is the process of making a square (or calculating its size) so that it is equal in area to the given shape, which is usually made of curves. *One of the famous problems of mathematics is the 'quadrature of the circle'; also known as* **squaring the circle**.

words translated into other languages

English	French	German
acute angle	angle aigu	spitzer Winkel
add up	additionner	addieren
algebra	algèbre	Algebra
angle	angle	Winkel
approximate	approximatif	approximativ
area	aire	Flächeninhalt
arithmetic	arithmétique	Arithmetik
average	moyenne	Durchschnitt
axis	axe	Achse
bar chart	barres graphique	Balkendiagramm
bisect	couper	halbieren
capacity	capacité	Kapazität
centre	centre	Zentrum
circle	cercle	Kreis
circumference	circonférence	Umfang
compound interest	intérêts composés	Zinseszins
concave	concave	konkav
cone	cône	Kegel
convex	convexe	konvex
coordinates	coordonnées	Koordinaten
cosine	cosinus	Kosinus
cube	cube	Würfel
cylinder	cylindre	Zylinder
decimal	décimal	dezimal
depreciation	dépréciation	Entwertung
diagonal	diagonal	Diagonale
diameter	diamètre	Durchmesser
divide	diviser	teilen
edge	tranchant	Rand
ellipse	ellipse	Ellipse
equal	égal	gleich
equation	équation	Gleichungformel
estimate	estimer	schätzen
even number	pair	gerade Zahl
factor	facteur	Faktor
formula	formule	Formel
fraction	fraction	Bruch
geometry	géométrie	Geometrie
graph	graphique	Graph
hexagon	hexagone	Sechseck
horizontal	horizontal	horizontal

hypotenuse	hypoténuse	Hypotenuse
infinity	infini	Unendlichkeit
isosceles	isocèle	gleichschenklig
mathematics	mathématiques	Mathematik
matrix	matrice	Matrix
negative	négatif	negativ
numbers	nombre	Zahl
obtuse angle	angle obtus	stumpfer Winkel
octagon	octagone	Achteck
odd number	impair	ungerade Zahl
parabola	parabole	Parabel
parallel	parallèle	Parallele
parallelogram	parallélogramme	Parallelogramm
pentagon	pentagone	Fünfeck
per cent	pour cent	Prozent
perimeter	périmètre	Begrenztheit, Peripherie
perpendicular	perpendiculaire	senkrecht
pie chart	graphique circulaire	Kreisdiagramm
plane	plan	Ebene
polygon	polygone	Polygon
polyhedron	polyèdre	Polyeder
positive	positif	positiv
prime number	nombre premier	Primzahl
radius	rayon	Radius
ratio	raison	Verhältnis
rhombus	losange	Rhombus
right angle	angle droit	rechter Winkel
sector	secteur	Kreisausschnitt
semicircle	demi-cercle	Halbkreis
series	série	Reihe, Folge
significant figures	chiffre significatif	geltende Stelle
sine	sinus	Sinus
square	carré	Quadrat
square root	racine carrée	Quadratwurzel
straight line	rectiligne	Gerade
subtract	soustraire	abziehen, subtrahieren
symmetry	symétrie	Symmetrie
tessellation	mosaique	Mosaik
tetrahedron	tétraèdre	Tetraeder
trapezium	trapèze	Trapez
whole number	nombre entier	ganze Zahl
vertex	sommet	Scheitel

words with their pronunciation

As a guide to the pronunciation of some words, they are broken up into parts with a 'spelling' much more like the sound to be made. So mathematics would look like math - uh - **mah** - tix. Parts printed **in bold** show where the stress, if any, should fall.

abscissa	ab - **siss** - ah
algorithm	**al** - gore - rythm
alidade	**ah** - lee - dayd
aliquot	ah - lee - **kwot**
amicable	ah - **mick** - ah - bull
aperiodic	**ay** - peer - ee - odd - ick
apex	**ay** - pecks
Archimedes	arky - **meed** - ees
arcsin	**ark** - sine
are *(measure)*	air
arithmetic	ah - **rith** - meh - tick
or	ah - rith - **meh** - tick
	(as in arithmetic mean etc.)
asterithms	**ass** - ter - rythms
astroid	**ass** - troyd
asymmetric	**ay** - sim - etrick
asymptote	**ass** - im - tote
axes	**acks** - ees
axis	**acks** - iss
bicimal	**bye** - sim - ull
bilateral	by - **lat** - err - ahl
binary	**bine** - ah - ree
binomial	**by** - nome - ee - al
bisection	by - **seck** - shun
bow compass	**boh** - cum - pass
cardioid	**car** - dee - oyd
catenary	kuh - **teen** - ah - ree
Celsius	**sell** - see - us
centroid	cen - **troyd**
chi	kye
chord	kord
clinometer	klin - **omm** - it - err
collinear	ko - **lin** - ee - uhr
commutative	komm - **you** - tuh - tiff
conjecture	kon - **jeck** - churr
crescent	**kress** - ent
cyclical	**sick** - lick- uhl
cycloid	**sye** - kloid
cyclotomy	sye - **klot** - uh - me
denary	**deen** - uh - ree
directrix	die - **reck** - trix
dissection	dis - **seck** - shun

Diophantine	die - uh - **fan** - teen
dodecahedron	doh - decka - **hee** - dron
domain	doh - **mane**
eccentric	eck - **sen** - trick
enumerate	ee - **new** - mer - ate
eponym	**ehp** - on - im
equiangular	eck - we - **ang** - you - lar
escribe	**ee** - scribe
Euclid	**you** - klid
Euler	**oil** - err
extrapolation	ex -**trap** - oh - lay - shun
fallacy	**fal** - uh - see
Fibonacci	fib - on - **ah** - chee
figurate	**fig** - ur - ate
focus	**foe** - cuss
foci	**foe** - sye
frieze	freez
Gauss	gowss
geodesic	jee - oh - **dee** - sick
gelosia	**jeh** - low - see - ah
geometric	gee - oh - **meh** - trick
giga	**geeg** - uh
gill	jill
googol	**goo** - goll
Graeco-Latin	**gree** - co - lah - tin
hexiamonds	hex - **eye** - uh - monds
Hypatia	hie - **pay** - sha
hypotenuse	high - **pot** - en - use
hypothesis	high - **poth** - uh - siss
hypotheses	high - **poth** - uh - sees
icosahedron	eye - cos - uh - **hee** - dron
indices	**in** - dis - ees
interpolation	in - **ter** - poh - **lay** - shun
intuitive	in - **tew** - it - iff
invalid	in - **val** - id
iota	eye - **oh** - tah
isometric	eye - soh - **meh** - trick
isosceles	eye - **soss** - uh - lees
iteration	it - err - **ay** - shun
kilogram	**kee** - loh - gram
kilometre	**kee** - loh - meet - err
Leibniz	**lie** - **b** - nits

lemma	**lem** - uh	Ptolemy	**tol** - uhm - ee
linear	**lin** - ee - uhr	quotient	**kwoh** - shunt
longitude	**lon** - ji - tewd	quotition	kwoh - **tish** - un
lune	loon	radii	**ray** - dee - eye
Mascheroni	mash - uh - **roan** - ee	Ramanujan	ram - uh - **new** - jan
matrices	**may** - tris - ees	reciprocal	**ree** - sip - roh - kuhl
matrix	**may** - trix	reductio	re - **duct** - ee - oh
median	**mee** - dee - an	repunits	re - **pun** - its
meridian	muh - **rid** - ee - an	residue	**rez** - id - you
Mersenne	**mare** - sen	rhombi	**rom** - bee
mnemonic	neh - **mon** - ick	scalene	**skay** - leen
Mobius	**moh** - be - us	Schlegel	**shlay** - gull
modulo	**mod** - you - loh	s'choty	**shot** - ee
mosaic	moze - **ay** - ick	secant	**see** - kant
motif	moh - **teef**	sequence	**see** - kwents
multiplicand	mull - **tip** - lih - cand	Soma	**soh** - muh
nephroid	**neff** - royd	soroban	**sorrow** - ban
node	noad	steradian	**stehr** - ay - dee - un
nominal	**nom** - in - ull	suan pan	**soo** - **an** - pan
nomogram	**nom** - oh - gram	tacheometry	tack - **ee** - om - uh - tree
nu	new	Tartaglia	tar - **tahl** - ee - uh
oblique	oh - **bleek**	tera	**teh** - ruh
ogive	**oh** - jive	ternary	**turn** - uh - ree
orthogonal	**orth** - ogg - uhn - ahl	tessellation	tess - uh - **lay** - shun
palindrome	**pah** - lin - drome	tetrahedron	**tet** - **ruh** - hee - dron
pandigital	pan - **didge** - it - uhl	Thales	**thay** - lees
parallelepiped	para - lel - uh - **pie** - ped	theorem	**thear** - um
parametric	para - **meh** - trick	topological	top - oh - **lodge** - ick - ull
parity	**pa** - rit - ee	transcendental	trans - en - **dent** - ull
partition	parr - **tish** - un	traversable	trah - **verse** - ah - bull
pedal	**peh** - duhl	trapezium	tra - **peeze** - ee - um
periodic	peer - ee - **odd** - ick	trapezoid	**trap** - eh - zoyd
periodicity	peer - ee - od - **diss** - itee	trigonometric	trig - on - uh - **met** - rick
permutation	perm - you - **tay** - shun	trinominal	try - **noh** - mee - ull
peta	**pee** - tuh	unary	**youn** - uh - ree
phi	fie	undecagon	un - **decka** - gon
pico	**pee** - co	unicursal	you - nee - **curse** - ull
planimeter	plah - **nim** - it - err	unique	**you** - neek
Platonic	plah - **tohn** - ick	union	**youn** - yun
polyiamonds	polly - **eye** - uh - monds	unitary	**youn** - it - arry
polyominoes	polly - **om** - in - ohs	valid	**vah** - lid
premise	**prem** - iss	variable	**vair** - ee - uh - bull
prism	**priz** - um	variance	**vair** - ee - ants
prismoid	priz - **moid**	vertex	**verr** - tex
pro rata	proh - **rah** - ta	vertices	**verr** - tiss - ees
psi	ps - eye	vinculum	**vin** - kew - lum

x must be the the the most familiar letter in mathematics. There can be few students who have not been faced at some time with a solution to a problem that starts, 'Let the unknown quantity be *x* ...' and the letters *y* and *z* follow soon after as the need for more unknowns increases. Why do we use those particular letters ?

The Ancient Egyptians (about 2000 BC) faced with a problem having only one unknown referred to it as 'the heap'.

Hindu mathematicians (600–700 AD) identified the first unknown as yã (short for 'yãvattãvat' meaning 'so much as') and any other unknowns in the same problem were named after the colours. Thus, the Hindu words for black, blue, yellow represented the 2nd, 3rd and 4th unknowns and so on.

Chinese mathematicians (1000–1200 AD) could only work with up to four unknowns. These were identified as 'the elements of heaven, earth, man and thing'. They could not go beyond four because their system required them to write each unknown in one of only four positions.

European mathematicians (1400–1500 AD) gradually took up the idea of using letters or other symbols to simplify their writing, but almost every writer favoured his own system and most of these systems were by no means straight forward, especially when it came to writing powers of the unknown quantity. In many cases the unknown was represented by one symbol, its square by another, its cube by another and so on.

Italian mathematicians (1300–1600 AD) developed a lot of algebra. Their unknown was referred to as 'cosa' (='thing') which led to algebra in England being described as 'the cossic art'. The square, cube and fourth power of the unknown were described as 'censo', 'cubo' and 'censo de censo' respectively, abbreviated to ce., cu. and ce.ce, but variations persisted.

x, x^2, x^3 was shown by one writer as *A, A quad, A cub,* which found some favour as it linked with the geometrical idea of x^2 representing area and x^3 being a volume. The English mathematician Thomas Harriot (1560–1621) wrote *a, aa, aaa,* etc to show the same thing.

René Descartes the French mathematician (1596–1650), produced a book in 1637 in which he proposed using the first letters of the alphabet to represent *known* quantities, and the last letters to represent *unknown* quantities. This led to his using *x* and *y* for the two axes in plane **Cartesian coordinates** which he also introduced. To this we add *z* when extending the system to 3 dimensions or when a 3rd unknown is needed. In his own work, if there was only one unknown in a problem then he always used *x*. His proposal was not adopted by everyone immediately, and even he felt it necessary to write to other mathematicians in their own systems. In a letter written in 1640 he used 1C – 6N = 40 to show the equation we would now write as $x^3 - 6x = 40$. However, his proposals did eventually become established and it is the convention we use today. He also proposed the form of **index notation** with which we are familiar.